FARMERS GUIDE TO ARABLE MACHINERY

B

FARMERS GUIDE TO ARABLE MACHINERY

HERMAN SIMPER

FARMING PRESS LIMITED
WHARFEDALE ROAD, IPSWICH, SUFFOLK

First published 1977

ISBN 0 85236 080 0

This book is set in 10 pt on 11 pt Times and is printed in Great Britain on Bowpaque
Offset Cartridge paper by Page Bros (Norwich) Ltd, Norwich

CONTENTS

CHAPTER PAGE

FOREWORD
by Rt Hon J. M. L. Prior, MP 12

PREFACE 13

ACKNOWLEDGMENTS 14

1 THE PRIME MOVER 15

The horse as draught animal—steam power—pioneers in machinery field—more owner-operators—cost of a tractor—the Gyrotiller—starting farming today—when buying a tractor—dealers and agencies—advantages of standardised machines—benefit of pneumatics—tendency towards four-wheel drive—the automatic tractor—no 'best buys' in farm machinery—choosing a tractor—hydrostatics the answer—Finland's bogie tractor—looking into the future—turbo-charging to stay—crawler tractors—follow line of market share

2 TILLAGE MACHINERY 43

Signs of breakthrough—round-the-field working—chisel ploughing—drill-and-till era?—deeper ploughing for beet and potatoes—no support for disc plough—the spring-tine cultivator—power-driven tillage tools—the rotary cultivator—the plough still supreme—effect of ploughing speed—the one-way plough—move to semi-mounted ploughs—disc harrows—ribbed-type and flat rolls—use of hydraulic power—the Roterra seedbed maker—the subsoiler—trailed and mounted types—farming without the plough

3 DRILLS, SPINNERS AND PLANTERS 66

Two spouts or one?—combine drilling—force-feed drills—rotary action and pneumatic types—popularity of tine drills—direct drilling—optimum space for higher yields—planting depth critical—uniformity of seed important—snags with precision drilling—see seed is well covered—disadvantage of individual hoppers—drill-and-till implement—self-propelled machines?—liquid fertilisers—potato planters

CHAPTER PAGE

4 FARM SPRAYERS 92

Instant weed elimination—wild oats: a serious problem—the mounted sprayer—lack of marker system—underlapping trouble—roller van pump—selecting nozzles—importance of boom height—mounted water tank—many sizes of sprayer—important trailer models—inaccurate marking—spraying by aircraft—which to buy?

5 THE COMBINE HARVESTER 107

Great mechanical advance—once-over operation—introduction of tanker combine—money problem—grain storage—swathing a non-starter—blowing better than rolling grain—the serrated knife a boon—pick-up reel favoured—butt-end-first-feed system—sealed silos—aeration desirable—judging right time to cut—straw inhibitors—combine prices justified—self-propulsion always a winner—successful makes—automatic height control—grain monitors—re-threshers—combining in hilly areas—roto-threshing: a Canadian concept—size and shape of cabs—round-and-round working best

6 THE STRAW HARVEST 134

Round bales—low-density balers—permanent cover for straw—burning: the economical way out—whole-crop harvesting—ram baler—picking up behind big combine—dealing with badly lodged crop—wafering—unrolling the bales—the bale sledge—random loading—endless chain pick-up conveyor—trailers designed for bales—processing straw with caustic soda—processing machines—future bright for balers—effect of climate—retail outlets an important factor

7 HARVESTING ROOTS 158

Drop beet production in UK?—sugarbeet: a good break crop—avoid beet on heavy land—early machinery developments—latest techniques—Continental harvesters—machines to use—the dirt problem—forage toppers—two types of cleaner-loader—revolving cage on front loader—the topper unit—beet harvester knives—flail toppers—potato harvesting—farm-to-consumer selling—judging a machine's value—testing prototypes—simple way to remove stones—self-propulsion and other developments—hydrostatic transmission—selecting machine for the job

8 FARM TRANSPORT 185

Tractor trailer combination—fitting full lighting board—the Land-Rover—old road trucks—whole idea: to reduce draught—concentrate on wheel traction—two-wheel tipping trailer—market for self-propelled transporter—mini container system—self-emptying trailers—containerisation—hydraulically-operated front-end loader—the weighbridge—summary

CHAPTER PAGE

9 BUYING MACHINERY AND TRAINING THE OPERATOR 200

Lease-backs—the farm contractor—short-term hire—NFU-sponsored syndicate schemes—secondhand machinery purchase—labour problems—training courses—courses run by manufacturers—information on micro-fiche—video display unit—tractor-cab comfort—communication difficulties—Continental machinery centres—teaching safety on the farm

APPENDIX 210

INDEX 213

ILLUSTRATIONS

		PAGE
1.	Early Ferguson tractor and 75 hp MF 185 tractor	17
2.	MF 1200 tractor and John Deere disc harrow on Dutch polders	18
3.	Cut-away of MF 1200 tractor	21
4.	International 37 hp and 374 tractor fitted with 'quiet cab'	21
5.	Ford 128 hp 8700 tractor fitted with four-wheel drive; plough is 6-furrow Dowdeswell	22
6.	72 hp Leyland 272 tractor with 4-cylinder engine and 10-speed gearbox	22
7.	215 hp John Deere 8430 tractor pulling levelling harrow	23
8.	Case 2670 pulling 12-furrow Kvernelands plough	23
9.	Big 'A' tractor from America with 110 hp engine	25
10.	British Marshall AM 105 crawler pulling a Lely Roterra	25
11.	County 1174, a typical four-wheel-drive tractor, pulling a set of Parminter disc harrows	27
12.	Inside cab of 1174 tractor	27
13.	Unimog tractor from Germany	28
14.	New Idea power unit	28
15.	A new concept in plough base design	51
16.	Ransomes Spaceframe model fitted with hydraulic reset trip legs	52
17.	Ransomes bar coupler cross shaft attached to tractor lower links	54
18.	With Ransomes C 92 cultivator, a basic frame can be extended as farmer increases the power of his tractors	57
19.	Extra-strong two-piece tines of Vibro Flex cultivator	57
20.	Lilliston rolling cultivator for inter-row work	58
21.	Lely Roterra	58
22.	Lely Combi 300 has 17 Suffolk-type coulters if fitted to Roterra 300	62
23.	Vicon four-bar harrow	62
24.	Bomford & Evershed Superflow, behind Track-Marshall 753 tractor	63
25.	Canadian Packer tractor	63
26.	Ransomes-Nordstein 3-metre cultivator drill	70
27.	Staahay Precision cereal drill	70
28.	Small version of Alpha-Accord drill	71
29.	Bettinson 3D direct-seeding drill	71
30.	Transporter for wide versions of modern drills	78
31.	6.7 m width of new 30-row model of MF 30 drill	78
32.	Stanhay 4-row S 870 precision seeder	79
33.	Howard Rotacaster	79
34.	Combined fertiliser-drill-spray operations with Deutz Intracs	82

PAGE
35. Faun two-row potato planter 82
36. Aerial picture of Chafer sprayer at work 86
37. Chafer's tramlining device 86
38. Ugerlose front-mounted hoe from Denmark 88
39. RDS Acremeter gives direct reading of acres worked 88
40. Everall 800L tractor-mounted agricultural sprayer 97
41. Ransomes 12-metre sprayer boom for metric sprayers—Cropguard 700 Super and 500 Super 97
42. Dorman Super TM4 mounted sprayer 100
43. Dorman 680 litre saddle-tank sprayer 100
44. Model 300 sprayer from E. Allman E Co Ltd fitted with boom winch 104
45. Allaeys self-propelled sprayer 104
46. Cut-away picture of MF 625 combine 120
47. Sperry New Holland Model 8080 giant combine harvester 120
48. The Claas Dominator 100 121
49. MF 760 combine has luxury cab pressured to kep it dust-free, with full air conditioning providing refrigerated or heated air as required 121
50. Danish JF combine built round David Brown tractor 122
51. John Deere 965 H Sidehill combine 122
52. New-type flat, dry-type sensor used in the RDS Mk 3 Monitor 125
53. Diagram showing positions of three full-width sensors of a new, dry pattern fitted in new RDS Mk 4 Monitor 125
54. Wilder hydraulic spreader kit, here fitted to a Wilder 1.83 metre straw chopper 130
55. Cab for International combine provides excellent all-round vision and can be equipped with air pressurisation unit. 130
56. Tanco bale loader 143
57. Gehl big baler at work 143
58. International 440 pick-up baler 144
59. McConnel Balepacker 144
60. How International 241 Bigroll baler forms a bale 145
61. Farmhand bale form used in conjunction with 8-bale accumulator 147
62. Farmhand Stackhand 147
63. Vicon Straw spreader at work, facilitating the subsequent ploughing-in of straw 150
64. Sperry New Holland automatic bale waggon 150
65. Forager from CTM Engineering loading pea haul from swath 152
66. Sperry New Holland Model 376 Hayliner baler in action 152
67. Armer-Salmon beet harvester 164
68. Standen Rapide Tanker 164
69. Standen Multibeet two-piece harvesting system 168
70. Standen Multibeet flail topper 168
71. Standen Solobeet Cyclone 169
72. Ransomes Powerbeet built round Ford tractor 169
73. Artist's drawing of Ransomes 2-row self-propelled beet harvester 170
74. CTM Super 120 sugarbeet cleaner/loader 170

PAGE

75. Herriau 3-part harvesting system from France 171
76. Whitsed Vulcan 2-row potato harvester at work 171
77. Diagram showing X-ray sorting system for Whitsed Automatic potato, vegetable and bulb harvester 176
78. Standen Startrite potato harvester in use with hand sorting 176
79. Grimme self-propelled potato harvester 177
80. Benedict Amac 2-row self-propelled potato and vegetable harvester 177
81. The Land-Rover 188
82. Steyr-Puch Halflinger 188
83. Rekord all-purpose vehicle 190
84. Combination involving forward control Land-Rover, new Holland forager and Farmhand body 190
85. Single axle chassis of the Krone Agro-Trailer 192
86. Carnfield 'La Baladeuse' high-lift trailer 192
87. Trailer constructed for bulk handling of seed and fertiliser from store to drill 194
88. 1-metre folding forks on McConnel Tail Slave TU20 fork lift 194
89. Conversion of standard Sanderson Power Fork 195
90. McConnel loader 195
91. Self-loading lime spreader, built on County forward control tractor, is fitted with McConnel Power Arm 44 loader 196

DIAGRAMS

Predicted life of world energy reserves 40
Creation of tram lines 85

FOREWORD

by Rt Hon J. M. L. Prior, MP

AGRICULTURE has undergone a revolution in the last 30 years. Nowhere is this more apparent than in the arable areas of East Anglia, and of course in Suffolk, where Herman Simper has gained so much of his vast practical and detailed knowledge. The scenery itself has changed. The fields are larger—in some cases unnecessarily so—and ridge and furrow have been replaced by one-way ploughing. Good drainage, always important, is now essential. The horse has given way to the tractor, and small, simple machinery to large, expensive and complicated apparatus. The weight of crops is much greater, the need for attention to detail more marked.

The pattern of crops has also changed. Vegetables, particularly peas, dwarf beans and brussels sprouts, require their own specialised machinery and contribute to widening the old four-course rotation.

Cereal growing, for so long the bread and butter of East Anglian farming, is under rapid change. Crop husbandry, the greater use of herbicides and insecticides, are combining with new varieties to produce a breakthrough of the 20 sacks per acre ambition of every farmer to yields unheard of five years ago. All these developments place enormous reliance on the development of new machinery and technical details.

This book is much more than a catalogue of different makes and types of farm machinery. It is a mine of information, both technical and practical. Narrated in a manner which reflects the author's long experience and detailed knowledge, it will help all those who seek by higher standards to serve the Earth and increase the supply of food to a world and country rapidly outgrowing in population the resources available to feed it.

JAMES PRIOR

House of Commons, Westminster
September, 1977

PREFACE

I HAVE not attempted to write a text book as there are one or two superb text books available on farm machinery. I have confined myself to field machinery in the arable sphere, because the forage section is already well covered. Nor have I delved into the specialised realm of such subjects as irrigation and grain-handling installations.

The book is written for the practical farmer, the young farmer and the beginner. Most chapters commence with a certain amount of historical evidence, because as background knowledge it can be useful in making decisions when purchasing machinery. They end with opinions that are expressed simply as guidelines.

My opinions are based on personal experience, the experiences of farming friends and also of people in the machinery trade and their demonstrators. The writing of this book was in fact generated by my thoughts on the lack of hard and fast opinion. I hope its contents will assist my fellow farmers in choosing their machinery wisely, especially in view of the high finance that is now involved.

I realise, of course, that it is impossible to mention every make of machine, and no disrespect is meant by the inevitable omissions. However, it is my hope that farm machinery manufacturers will keep me informed of modifications, improvements and new introductions, so that I may be able to re-assess machines for any future revision of the book. I apologise for the incontestable fact that some machines may be modified or superseded between the time of writing and the publication date.

I ask my friends in the trade to accept my opinions as a challenge to the production of better and more reliable machinery.

Machinery is now an everyday part of farming life, and although it may not be possible to test machines on a par with motor cars, the need for independent testing and individual opinion is just as important. I have tried to fill the gap that exists immediately behind the text book study arena, but I cannot over-stress that my stated preference for any particular machine has to be balanced with local circumstances and local agencies. Not least, the individual selection of a machine has to be balanced against the fiscal policy in force at any particular time.

I have deliberately interspersed opinion with technical description of machines in most chapters, because it so happened to fit in with the section of machinery under discussion.

In most cases, but not all, I have ignored price when giving my choice. For one thing, prices are continually changing and also many farmers are not unduly parsimonious regarding their mechanical fancy. In a few cases I have gone out of my way to mention a value-for-money buy.

I have tried to project my thoughts ahead with prophesies of future machinery in some chapters. For the same reason I have sometimes mentioned machines that are uncommon and may not even see the commercial light of day. But, mainly, I have attempted to show the reader what is on the market in the arable field and what my choice would be, but the over-riding message is that every farmer's decision has to be influenced by the service that is available in the area in which he farms.

H. SIMPER

Pulham Market, Norfolk
March, 1977

ACKNOWLEDGMENTS

I WISH to thank my many friends in the public relations departments of the following firms who have supplied information or pictures: Allmans, Bamfords, Bomford & Evershed, Benedict, Bunns, Chafers, David Brown, Dormans, Farmhand, Fisons, Ford, Howard Machinery, International Harvester Company, John Deere, J.F. Ltd; Lely Imports, Leyland, Manns, Massey Ferguson, Ransomes, Richard Pearson, Root Harvesters, Sperry New Holland, Standens, Stanhays, Vicon.

Thanks are also due to the Agricultural Training Board, Agripress Publicity, British Sugar Corporation, Four Seasons Publicity, Ministry of Agriculture, National Institute of Agricutural Engineering, Potato Marketing Board, The National Farmers Union.

I am also grateful to Mrs Phyllis Durbidge for her valuable typing assistance.

H.S.

Chapter 1

THE PRIME MOVER

LIKE all farm vehicles, the farm tractor is a means to an end. It is not really important whether it has a wheel at each corner, is round, square or oblong, nor whether it is painted red, white or blue. It happens to be the agricultural prime mover of our time and is capable of a power conversion that makes it obviously a more economical power unit than a human mustering a feeble one-eighth of a horse-power.

Before we become engrossed in the technicalities of the sophisticated giants that now adorn our countryside, it is as well to remember that our fathers, grandfathers and forefathers were all harshly limited in the power available to them. Not for them a saunter into the High Street, a train to London for Smithfield week or a visit to a multi-machine Royal Show display.

THE HORSE AS A DRAUGHT ANIMAL

Primitive man managed to harness the ox, the ass, the mule and the horse. So well did the horse perform, that just 100 years ago our grandfathers possessed about 850,000 of these noble creatures. Around 1910, there were almost one million, but in 1950 the tractor numbers overtook the horse.

A horse could well be consuming five acres of produce off the farm. Maybe the good farmer on good land could get away with three or four acres apiece for his draught animals. In all, these members of the equine species could well have been devouring the produce from 10 million acres.

Any good horseman will tell you that the essence of the art is not to let the animal know that you have any fear of it. Tractors have to be treated likewise. Whether they have inherited prejudices from the costly horse, I do not know, but there is a similarity. This metallic hulk of tractor seems to embody an innate knowledge that tends to cause it to tease the learner-driver.

STEAM POWER

Before the tractor arrived on our farms, we had a mild flirtation with steam power. We produced our own coal and the supply was well nigh inexhaustible.

The steam engine, one at each end of the field hauling its tackle to and fro, did even less damage to the soil than the horse hoof. But it did not take many years to find the snags and the answers. Steam, petrol, paraffin, tractor vaporising oil, then diesel—we went through all the fuels. The steam engine's appetite for fuel and water, plus the labour involved, made these ungainly monsters non-starters. Petrol was a satisfactory fuel, but every Chancellor of the Exchequer found it such a convenient source of revenue that it was soon to be taxed out of the field.

Tractors did not relish paraffin, TVO became the runner, with petrol the starter. Switching from one to the other with standard Fordsons in the 1930s was a work of art. One's judgment had to be supreme, or you had to start all over again, invariably invoking the wrath of the boss who was paying for the petrol.

PIONEERS IN MACHINERY FIELD

Tractors had started in America, and the Otto engine of four-stroke cycle had been introduced in Germany, but surely Mr Diesel and the compression engine in 1892 must have been the best breakthrough for farm machinery. We are all indebted to Henry Ford, and later Harry Ferguson, for true greatness in engineering design, mass production and marketing techniques.

As the horses went, so did the men. It was no longer necessary to be up at five in the morning to bait the prime mover. The combination on a one-eighth horsepower human and a full horsepower on four well-shod feet had served the farming community reasonably well in their ignorance. And what ignorance! Use man's miserable bit of power indirectly, via drawbar, PTO and later, hydraulics, and the cost per horsepower hour can be reduced to a mere fraction of the figure we first thought of.

From over half a million farmworkers in 1939, we got down to nearly half that number by 1969. During the same period, tractors rose from 55,000 to around 350,000, but the real change from man-power to horsepower came in the years between 1960 and 1964. The horse population had dropped from the astronomical 968,000 in 1910 to 46,000 in 1960, when we virtually wrote them off as of little further use. It was the 1964 figures that caused the big turnabout,

1 The introduction of the 75 hp MF 185 tractor (lower photo) in 1971 marked 25 years of Ferguson System farming in the UK. The popular little grey 'Fergie' (above) went into large-scale production in Britain in 1946. More than 500,000 of these tractors were sold throughout the world.

2 The author (left) discussing the work of a MF 1200 and a large John Deere disc harrow on the Dutch Polders.

when tractors exceeded working humans in farming. Whereas workers had come down from 511,000 in 1939 to 333,000 in 1964, tractors had soared to 389,000. Since that year, tractors had got thinner on the ground but steadily more powerful.

MORE OWNER-OPERATORS

It is not irrelevant to point out that the same sort of thing occurred in the figures relating to farmers and the whole-time workers. By 1973 there were fewer workers than farmers—about 35,000 less—and now (1977) there are around 40,000 less. The mechanical significance of this is that it means more and more owner-drivers of farm machinery.

Although farms have got steadily larger, the tractors more powerful and the labour force weaker, even now the average farm size is only 45 hectares. By removing a man from the drudgery of pick or shovel, it is easy to reduce his horsepower per hour cost to one per cent of his muscle cost. Obviously, anyone can go a long way with that sort of theory by adding more and more horsepower to the tractor.

On the other hand, farmers can make their own compensatory factors. The farmer's time is his own, as it were. Where the modern worker will not tolerate drudgery or outdated tools, it could well be that many working farmers will. The small farmer or, more likely, the beginner will compensate for under-powered tractors by working all the day and half the night. Later on, when established, he will want the sophisticated new generation of tractors.

I am certain that on the larger farms it is the operators, not the bosses, who demand these new, built-in comfort types. I doubt whether the demand for luxury and comfort in tractors comes from the owner-operators as the tractor makers seem to suppose. Much of the demand, in any case, comes from a simple matter of law. The safety cab, the safety frame, and now the below 90-decibel cab, have inevitably stimulated both interest and demand.

Anyone looking on from the outside might well be conned into believing that British farmers buy nothing less than 100 hp tractors. But in 1975 80 per cent of tractors sold were under 80 hp, and the average is still only 47 hp. It is predicted that horsepower will increase to an average of 60 by 1980 and to 70 by 1985, with the present figure of 15 per cent over 80 hp and 5 per cent over 100 hp going up to one quarter over 80 hp by 1980, and two-thirds over 80 hp by 1985, with half of them over 100 hp.

COST OF A TRACTOR

The historical evidence is that the average power unit of the farm is always going to cost about what is needed to pay, house and nurture the farmworker for a year. The standard Fordson that was costing £120 in 1927, had risen to £180 by 1937 and to £248 by 1947. By 1957 it was costing £598 and by 1970 it reached the alarming price of £1648. But by 1957 our common power unit had gone diesel, and horsepower that started at 23 had gone up to 40. The 1970 model that had superseded the 'Major' not only had the modern, compression engine but was now called 'Ford 5000', with 75 horses under its bonnet and the operator protected with a cab. Price per horsepower had crept up from £5 to £22. What reckless inflation! Today it is no trouble to find a crawler or four-wheel drive tractor that will set the farmer back £100 per horsepower.

Never forget that the price change between 1927 and 1937 took into account the change from steels to pneumatics! At Michaelmas 1946 the farmworker's wage was £4.00 per week. Tables show the similarity between what the farmer is prepared to pay for his prime mover and what he pays a man over a year. Similar tables show that the farmer has a remarkable habit of spending around 20 per cent of his annual net income on machinery.

When we refer to farmers buying machinery, inevitably we are alluding primarily to the farm tractor. Every arable farm, or predominantly arable holding, has a system of mechanisation that revolves round the common agricultural prime mover of the age. Its model number, age, type, colour and whether bought new or secondhand, is somewhat irrelevant to the issue. It hauls, it pushes and with loader arms it will lift. We started by taking the horse out of the shafts and putting the tractor in. In those early days some 90 per cent of these prime movers were the standard Fordson, similar to those that had come across the Atlantic to help out in World War 1.

These were aided and abetted by a spattering of International Farmalls, the John Deere 'D', Allis-Chalmers 'U', Lanz Bulldog, Minneapolis Moline, the Black Ferguson, the Ferguson-Brown, the Ford Ferguson, Oliver, Case, Massey Harris, Fowler, Rushton, Marshall, Clayton, Weeks, Austins, Ivels and Saundersons. Some of these old tractors, the Case for instance, were superb at belt pulley work. Remember, in the 1920s and 1930s the threshing tackle-was much in vogue, and as steam power dwindled, some of the old-stagers took over the belt drive from prime mover to drum.

Historical facts relating to the tractor illustrate just how conservative we have been. Talk to anyone from Industry or the motor-car sphere and they are likely to say that the tractor is a thing with a wheel at each corner and are likely to ask what more can anyone do to it. Quite a lot, as it happens. A look back also serves to remind us of what might have been. Either farmers lacked courage or, more likely, funds. Those of us brought up on family farms know that it was mainly the latter. How easily one can nostalgically look at the 1930 Massey Harris four-equal-wheel-drive model and breathe a heavy sigh. Replace steel with rubber and stick a cab on it and it would have a nice balanced look, even today.

THE GYROTILLER

One of the greats that got away was that first self-propelled tillage machine known as the Gyrotiller, that arrived on the scene in 1933. Anyone from the old Fowler company who looks at such implements as Lely's Roterra must smugly claim that he thought of it first. Maybe Fowlers simply put the Gyrotiller tines in too deep, but it still took farmers the best part of 40 years to really catch on.

3 This cut-away of the MF 1200 shows main points of a centre pivot, or articulated tractor.

4 The International 374 of 37 horsepower, fitted with a 'quiet cab', was introduced as an economical, all-purpose, runabout tractor.

5 The Ford 128 hp 8700 tractor fitted with four-wheel drive which can be hydraulically engaged on the move. The plough is a 6-furrow Dowdeswell.

6 The 72 hp Leyland 272. It has a 4-cylinder engine, 10-speed gearbox, independent pto, independent hydraulics, hydrostatic power steering and the Leyland 'Q' cab.

7 The 215 hp John Deere 8430 pulling a levelling harrow at around 8 hectares per hour. The Haylock 9.25 metres harrow is one of the largest of its type.

8 The Case 2670 from America is marketed by their associate company, David Brown. Here it is pulling the 12-furrow Kvernelands plough from Norway, marketed in the UK by Bamfords.

Just as the Roterra is one of the great cultivating inventions of our time, so could the Gyrotiller have been in 1933. We can curse the Industrial Revolution, cheap grains from the Western Hemisphere, urban-orientated members of Parliament, a weak National Farmers Union, or world recession. Whatever the cause and effect, farmers were held up. So the Gyrotiller got away, only to rust and rot whilst politicians went on a spending spree to produce groundnuts with Gyrotillers in far-off lands.

At the Verona Fair in the mid-70s, I saw crowds of enquiring agriculturalists doing various contortions and knee-bend exercises to fathom the skill and mystery embodied in the Cantone self-propelled tillage machine. I stood back and wondered whether many, or any, of them had seen or heard of the British masterpiece that was Fowlers Gyrotiller. True it only cultivated, but the complete cultivate-and-seed-and-fertilise machine would have emerged. Alas, it looks like being the best part of half a century before we go back to take a second look.

The historic preamble is important. Many youngsters I find, harp after something that has been rejected, damned and annihilated. Others do not even know what has been and gone. This is not a matter for alarm or despair, because the average youngster has an enquiring mind that could cause the resurrection of the good ones that got away. It is unlikely, however regrettable, that we shall see any more Henry Fords or Harry Fergusons, because of society changes from individuality to institutional procedures.

STARTING FARMING TODAY

The average young man who goes farming is almost certainly starting on a tight budget. Not as meagre as many of us had to start with, as plenty of sons today are merged into their father's already thriving farming businesses. The spade work was a generation back. Despite the alarming stories about finance required, lack of land, lack of tenancies and lack of incentive in an urbanised economy many young men are still willing to have a go. What is more, many succeed. Like the people who come into farming from other industries, often armed with a useful golden-handshake, some fall by the wayside, some survive and others become shining examples for all to see.

The first thing to buy when anyone goes farming is the prime mover. Like it or not, that is the system. Self-propulsion is confined to farming spheres, mainly harvesting techniques with the combine harvester. The tractor is still a maid of all work, taken out of one lot of shafts, as it were, and put into a different cart.

9 The Big 'A' from America, with its 210 hp Detroit engine, is used for contractor work on fertiliser spreading and similar jobs. A spray tank is available and the giant tyres have flotation properties ideal for difficult conditions.

10 The British Marshall AM105 crawler is here seen pulling a Lely Roterra at a demonstration.

At the outset it should be said that the average modern British-made diesel tractor is a fairly reliable piece of equipment. Not as foolproof as we would like and lacking in fine arts, but reasonable. Also, there is little to choose between most of the main breeds of tractor either on specification or price.

This is the inevitable result of our stereotyped society. But to be fair to tractor manufacturers, they are no worse than any other branch of industry and better than some. Certainly their after-sales service, their urgency in repair work, their willingness to truck and trade, and the spirit of dutifulness over spare parts, leaves the motor car industry a poor second.

WHEN BUYING A TRACTOR

Which is the best tractor buy? Which one performs best? Which is cheapest? Can you buy this one or that one and get service? In my early days in farming, I asked a large farmer which was the best make of combine. He replied that it is the one with the best man on the seat! The phrase could equally apply to a motor car, combine or any other machinery propelled by the internal combustion engine.

I suggest that, using a car with a trip mileage recorder showing tenths of a mile, you measure the distance of local agents from your farm. Having done that, the logical answer would be to buy from the nearest depot. That would be all right if all agents were alike, but they are not. Thus the individual has to assess the service and spares availability against the mileage involved. Never forget that mileage costs work both ways. It is not just a matter of your own petrol and time, but the mileage cost of a dealer's vehicle, possibly with two men on board, travelling those miles for each service and breakdown.

As I have already said, there is not a great deal of difference between the main makes of tractor, and any variation in price is often only temporary until the rest of the field catch up! Nor, believe it or not, are the number of retail outlets for tractors in proportion to the market share. All the principal market contenders are very similar in retail strength.

DEALERS AND AGENCIES

Fords have 122 dealers and 43 agents, making a total of 165 retail outlets for their tractors.

Massey Ferguson have 81 distributors but they control 182 distribution depots. Add 26 dealers and you have a total of 208 premises where terms can be argued.

11 The County 1174 is typical of the four-wheel-drive tractor that many heavy land farmers buy for ploughing and heavy work. Here it is pulling a large set of Parminter disc harrows. **12** Inside the cab of the 1174.

13 The Unimog from Germany is equally at home ploughing, transporting or being driven into town.

14 The New Idea power unit has use as a portable powerhouse for propelling foragers, combines and so on.

International Harvester have a total of 154 places where their products can be purchased direct.

David Brown boast 205 sales agencies for their tractor range.

Leyland make claim to 80 distributors and 80 dealers, thus making a total of 160 outlets.

John Deere have 77 sales bases for their machines, 64 of them are in England compared with the whole of the United Kingdom.

Fiat and Zetor, to pick two from the foreign fields, have 39 and 50 sales points respectively in this country. Fiat have 30 agents in England and another 10 who deal only in crawlers, thus making a grand total of 40. Of the Zetor agencies, 35 are in England.

Massey Ferguson have always claimed that a farmer can find a depot within 25 miles of home; exceptions would be the North and West of Scotland. David Brown have enlarged their network of depots considerably over the past 10 years or so. International aim to have depots spaced about every 30 miles; again, this excludes the marginal areas. While these figures may change slightly and temporarily, it is not by very much, for the simple reason that if an agent goes out of business, the manufacturer will quickly be looking for a replacement.

Most farms are badly and inefficiently mechanised. Walk into a farmyard and see a conglomeration of different tractor makes and it is easy to decide that here we have a cast-iron case of wasteful expenditure. Often this comes about for no other reason than extracting the last penny in a trade-in. Equally, it can happen if the farmer buys the one that came out on top on a test report, a demonstration, or a competition, public or private.

When it comes to buying machinery very few farmers have the courage of their own convictions. Only one in a hundred would do a package deal for five or six tractors of the same make and model. Mechanical wastefulness comes not only from an assortment of breeds, but also from a bevy of models. Thus we think that it is necessary to have a large tractor, a small one and one in the middle.

The disadvantage is weight. We do not have to use the built-in horsepower, but we have to tolerate the weight that goes with power. For ploughing and cultivating jobs it may be immaterial. For row-crop work, spraying and fertilising the weight can be harmful. Again, it is a matter of getting priorities right and striking the correct balance. All this has to be weighed, sifted and analysed in the cause of standardisation.

Nobody starting arable farming should have more than two manufacturers' models on the farm and preferably only one. Unfortunately, because of the need for volume production to cut costs the 'full-line' concept in farm machinery has been partly abandoned. In the UK only Massey Ferguson got near to it. Nowadays it is more common for a successful tool manufacturer, here or overseas, to come to a manufacturing arrangement with an equally successful tractor maker.

Tractors have become more complicated, more sophisticated and larger, so there may be some excuse for the conglomeration but I deplore it. Any apprentice can see where nut and bolt sizes could be brought down to manageable proportions. The dismal and inexcusable fact is that most tractor manufacturers, having duplicated sizes beyond reason, send their tractors to the farm with toolboxes filled with fresh air! If there are any tools, the likelihood is that the tractor is imported.

ADVANTAGE OF STANDARDISED MACHINES

As with so many other things in this vast machinery world, benefits can bring their own troubles. There is a tremendous operator advantage in standardised machines in that the staff will be conversant with all the tractors. Yet standard sized spanners can leap from toolbox to toolbox! It also applies to any moveable part of the prime mover. One way to prevent losses is to paint tools, top link, drawbar, filler caps and so on, with a different colour for each tractor. Another way is to file marks across all such appendages to correspond with the tractor number. A small but practical point.

One man, one tractor, is a dictum that serves farmers well. That way the maintenance is likely to be better and the smash-ups less. Yet there is the valuable advantage that any man can operate his fellow worker's machine in emergency, during holidays or during sickness. I suppose that one of the most important aspects of the standardisation policy is in relation to spares and repairs.

SPARE WHEEL A 'MUST'

The spare wheel, in this the age of pneumatics and punctures, is a 'must'. With all models alike, one front and one rear wheel as spares can save an almost incalculable amount of money over the four seasons.

The spare-wheel system is wonderful, but it is not a reason for complacency. Being too casual in getting the repair job done, or having men who do not report that they have used the spare, can become as expensive as on-the-field service. Flat spares are useless, as every motorist knows. It applies equally to the farm tractor spares.

A still newer system has recently come into vogue. Twin wheels are far from new, either on the road or in agriculture, but there has been a definite trend towards them. Thus we have our own built-in spare wheel system, so far as the rear wheels are concerned. A tractor can temporarily run with a wheel missing, or a wheel can be transferred to another tractor. Obviously, the twin-wheel system is becoming popular because of wheel adhesion difficulties in modern, high-powered tractors and because of excessive weight. Cage wheels do not escape from this reckoning. To my way of thinking, cage wheels always had that temporary Heath-Robinson look about them. They serve a good purpose at a reasonable price, but metal versus rubber is not a good argument.

BENEFIT OF PNEUMATICS

Pneumatics, like hydraulics, have increased the versatility of tractors. We may only get an average field efficiency of about 60 per cent out of our tractor tyres, but nobody wants to go back to steel wheels. Four-wheel-drive tractors would increase tyre efficiency to around 70 per cent. New tyres of the open tread pattern, biting full depth into friable, heavy land, look good and are reasonably good, but the average wheel slip is commonly put at 15 per cent.

The bigger the tyre lug, the more traction—at least, for the most part, that is so. In deep, soft soils the theory works, but on very dry, loose soil a smooth tyre might do as well. Also, it has to be remembered that the deeper the lug, the more rolling resistance to absorb power. Anything more than 50 mm lug height could be more trouble than it is worth. Radial tyres may give up to 7 per cent more grip than cross-ply, but they cost more and may increase rolling resistance. The advantages of radial against cross-ply would be likely to rise proportionately, on a par with road against land work. The flexible sidewalls give greater ground contact with radial tyres with the obvious results, but they do not surpass traditional tyres by such a wide margin as they do on road vehicles.

Often a radial tyre will give about the same performance as a cross-ply with water ballast. The convenience factor means that this is a worthwhile alternative. Ballasted tyres are a nuisance when a puncture occurs, especially if a ballasting kit is not available on the farm. It is easy enough to ballast tyres with tap water, but the adding of anti-freeze agents in the form of sodium chlorate or salt is not easy. However, it is a wise precaution as the ice can cause punctures.

It is safest to keep all tractors on an anti-freeze mixture so far as radiators are concerned. The easy way is to buy the common 25 per cent mix raw material, pour into suitable receptacle, fill the container that the anti-freeze came from three times with water, and add. Modern tractors use little water. It is not a great expense to top up radiators with the mixture the whole year round.

If, on the other hand, summer topping up has been done with water obviously, cooling systems have to be drained around November time and refilled with a suitable anti-freeze mixture. Air-cooled tractors are not as common as either air-cooled cars or sealed cooling system cars. Deutz go for air cooling, however, and make a satisfactory job of it.

TENDENCY TOWARDS FOUR-WHEEL DRIVE

Looking ahead, and remembering that many facets of machine development turn a full cycle, many people believe that, just as the Otto four-stroke cycle got going in 1864, so the Deutz Intrac system might get off the ground in the 1980s. There had to be a tendency towards four-wheel drive as tractor power increased. For the immediate future we are on Harry Ferguson's three-point linkage pick-up at the rear only. Like Mercedes-Benz, Deutz try hard to re-educate us with a new configuration, where implements can be attached, via three-point linkage, at either front or rear, or both.

Mindful of the manual struggle to line up modern, heavy cultivating implements, if hitching has to be done on an uneven surface, the Intrac is adept at automatic picking up. Not that such automatic pick-up frames are barred from farmers with ordinary tractors. They can be bought as extras, and two-point picking up on the lower link arms is available, with the operator manually dealing with the top link. The balanced, four-wheel drive layout of such tractors as Mercedes Benz, Deutz, Schilter and also the Fendt air-cooled diesel tool carrier do serve their purpose of making farmers think ahead.

The tool carrier concept has not caught on in the UK. Such wonderful bygones as David Brown's 2D are examples of machine, method and application being so logical and so right, yet spurned by the farming community. The profit margin on arable crops has not allowed for specialised tractors that work while their counterparts stand still. Thus the farm tractor has to be ubiquitous. For the future, if the small tool carrier type is not a starter because it only supplements its bigger brother, then farmers could look at a big-powered, all-purpose tractor of the Deutz layout. The alternative is to stay exactly as we are, with the tractor performing superbly at basic cultivating jobs as originally designed for, but more than a little clumsily for precision work.

THE AUTOMATIC TRACTOR

For repetitive, straight up-and-down work, there would be scope for the automatic or programmed tractor. Such tractors are suitable for work between fruit trees and indeed are being used on a limited scale for this sort of work. For large-scale field operation, where the ancilliary equipment can easily choke, break, or falter, these principles of automation are less attractive. From the technical point of view, it is easy enough to have failsafe arrangements, like the cow-catcher rope on automatic horticultural tractors, but a tractor that is safe-failed and hors de combat can be expensive luxury. It is thus extremely unlikely that the immediate future holds out much hope for complete automation with no operator on the seat.

Tractor production is big business, even by twentieth century standards. It is no longer a case of the great pioneer inventor, risking his meagre pennies and dreaming of fame and fortune at the end of the line. It is now a matter of a consensus of opinion at board-room level and the need to placate shareholders large, medium and small, and possibly in many different territories.

It is evolution rather than revolution. A steady plod of progress as against spectacular happenings. The case for drastic change in configuration has to be cast-iron and this can never be. Even the differential lock, now almost standard on tractors, has kept many farmers on two-wheel drive, where the four-wheel drive, with its additional expense, could have become standard.

NO 'BEST BUYS' IN FARM MACHINERY

There is rarely a best buy in agricultural machinery. The market is so competitive and the agencies so similar that, in some circumstances, the farmer could almost toss a coin for his decision. In fact, the decision to buy will be more likely influenced by the distance between farm and agent.

Occasionally, a particular model does dominate in its class and gather some sort of knight-errant image. But they can be counted on the fingers. The first was the standard Ford, mainly because it had virtually no competition. Then, of course, the legendary Ferguson TE20. The David Brown 990 carved a niche in the hearts of many farmers, as did the little International 275 as a loader tractor. The Massey Ferguson 135 seemed to assume leadership in its class at one time. Then the Ford 5000, and the Ford 7000, seemed to enjoy a lordliness as the ploughman's tractor. Either of the latter two, almost certainly pulling a Ransomes TS84 plough, can be seen enjoying a supremacy that many would envy.

Many manufacturers would like to be the Ford or the Massey Ferguson of Britain. Ironically, these company's main shareholdings are respectively in Detroit and Toronto. These two dominate the UK market and it has to be said that if trade-in value in any part of the British Isles is the criterion, then one's tractors have to be one or the other. Great competitors that they are, their models will always be similar and prices never far apart.

With two firms sharing some 60 per cent of the UK market, the competitors have to be pretty good to even stay in the race. Most manufacturers go for lower link sensing on the really big ones and for turbo charging to get the last ounce out of some of the engines on the upper-middle-class tractors, as well as the giants.

Getting farmworkers to put in, say 1000–1500 hours a year to justify the cost of the prime mover is not easy. The average driver on the average farm is likely to only want to clock up about 600 hours per year. With contractors it can be different. The farmer probably tolerates the idiosyncracies of his workers with a lighter heart than can the contractor. The latter cannot afford men who want to do less than 1000 hours and more likely will want 1500–2000 hours worked.

Since the late 1960s I have been well satisfied with Ford tractors. They are reasonably simple, a beefy back-end copes with the power from the engine, and they are not advanced beyond the talents of the men who use them. Ford agents are usually old-established firms, solid, dependable and practical. It is a tribute to Ford's well-known zeal for service and spares that so many firms want to use their skid units. Examples are Muir-Hill, Roadless and County.

CHOOSING A TRACTOR

Tractors are rather like horses for courses. For light land, if one wanted a ton-up monster, one would go for the American-type 2WD. For heavy land, one would plump for a County, Roadless, or Muir-Hill. The latter, 4WD types, are also suitable for what would be termed medium land. One big advantage with these types is Ford's excellent after-sales service and spares availability, because most of these tractors get marketed through Ford agents.

A question often asked is whether to have equal or unequal wheel layout. Provided the extra pennies are not too much of a problem. I would recommend four big wheels. The smaller farmer so often claims that there is scope for a no-frills, utility sort of tractor. Apart from the export trade to developing countries, I doubt it, only because the saving either to manufacturer in the making, or farmer in the using, may not be all that large.

Nonetheless, there is a modicum of sense in the argument, and just as Massey Ferguson left the 135 on the assembly line when they announced the 500 series, so International pander slightly to this theory with the 374. They call it an all-purpose runabout of 37 hp. A small farmer can buy a useful secondhand tractor for about the same money and get more power and more sophistication. But if he insists on a new machine, then it is no good saying that nothing is available. There is no 'cheap' tractor, for the simple reason that this is no longer possible unless you build from a do-it-yourself kit and do not value your time. Perhaps such a kit will become available some day.

Those who were reared during the early days of the McCormick binder and the early Farmalls will harbour a soft spot for all things International. The Chicago-based International Harvester Company can hardly drop the 'Harvester' while people remember the name of Semus McCormick! If one had to pick a range of tractors that the UK farmer tend to under-estimate, this would be it.

INTERNATIONAL HARVESTER'S INNOVATION

In the cab sphere International stole a march. Unlike their earlier record IH, so often first in the field, then submerged in the pack, introduced their Iso-mounted cab with rubber insulation and Sundyn heat-rejecting glass. They anticipated the June 1976 laws and marketed futuristic tractors at a price that made good economic sense. Farmers responded appropriately. The market share of IH rocketed up from a springboard of a safety law and this was something that no legislator could have foreseen!

Although Massey Ferguson have entered the crawler market in a small way in Britain, IH offer the most complete and all-embracing range of tractors to the UK market. Farmer or contractor, however large and varied the work call, now find it is possible to standardise on IH if they so wish.

I can give no logical explanation except colour, but if wives and daughters choose the tractors, it is likely that they would all be David Browns! IH and Brown (now merged with Case) rather vie with each other for 'firsts', not all of which stay the course. The autodrive Automatic gearbox, introduced by DB at the 1961 Smithfield Show, was not commercially produced because market research indicated that farmers were not willing to pay the extra £250. In 1971 DB modified their ideas to semi-automatic Hydra

Shift, which got a silver medal from the RASE in 1972, a Queen's Award in 1974 and a Design Council Award in 1976.

IH, alone amongst our manufacturers, offer the farmer hydrostatic transmission if he so desires. Browns did not get too much interest when they tried to do away with gear levers in 1961, nor did IH in 1970 when they introduced hydrostatics. This is not to say that this will always be the case. One rather defeats the other, so it would be foolish to predict that both systems will come.

HYDROSTATICS THE ANSWER

However good and useful hydra-shift, multi-power and dual power is on the various tractor ranges, one cannot help thinking that hydrostatics has to be the answer. Eventually this could be the standardised transmission system. Certainly something simple, such as the single-lever control system, has to eventually replace the multiplicity of gears and levers. The torque converter has not caught on with the agricultural tractor in the same fashion as with the industrial tractor.

Many torque converters find their way into farming via the secondhand market, following the initial depreciation in industry where cost-plus calculations can accommodate more sophistication. It is the fore-and-aft work of using a front-end loader where the torque converter really comes into its own. But it cannot be overlooked that full hydrostatic transmission would supersede all the half-way type systems that have been tried.

Returning to the theme of firsts, while Ford and Massey Ferguson have their own records of new introductions, those challenging from a position immediately below them are so often the innovators. In the same year that they introduced hydrostatics, International brought in the full synchromesh transmission. IH claim to be the first to offer differential lock as standard on a regular production basis in 1955 and disc brakes in 1953. Their approved safety cab goes back to 1966. David Brown's pioneering gets mixed up in the Harry Ferguson story, as is well known, but their lift and linkage goes back to 1937. The two-speed pto has its roots back in 1948.

Brown's high-speed direct injection engines came in 1949 and traction control in 1953. Their approved cabs were two years after IH in 1968 but on all models.

International and David Brown are the two main runners-up to Ford and Massey Ferguson in the market to supply the farmer with his prime mover. Leyland's tractor operations really only started in 1969, following the Nuffield's that had a famous and respected name but did not really penetrate. I doubt whether we can claim that the introduction of a Mini tractor of 25 hp, to follow in the wake of a car name that had become a household word, was a rip-roaring success.

With hydraulics improved and the tidying-up done, not only are the modern Leylands competitive, despite a none-too-large market share, but their performance at competitive events has been a very pleasant surprise.

One modern guessing game is to estimate the top horsepower figure that our farm tractors will reach. Every prophet in the business has to continually uprate the figure! The truth is that none of us knows. Figures have to be altered according to farming fashions in power requirements and to align with farm size.

300 HORSEPOWER?

Obviously, a 40-hectare farmer does not need a 150 hp tractor unless for some specific purpose peculiar to his farm. As the average size of farms increase, horsepower will increase pro rata for the obvious reasons, though tractor numbers would be likely to decline. At present (1977) it seems that a top of 300 hp is about right, but this can alter both ways. It should not be overlooked that if direct seeding was eventually proclaimed a success by the farming community, so that we no longer turned over 2,500 tonnes of earth per hectare to grow crops, the farmer may find that he has not got use for these high-sounding horsepower figures. Either that or many fewer power units would be required.

We already have the huge White tractors from America. In 1976 I saw a White being tried out on Holland's polder lands, where the Massey Ferguson 1200 is popular. If tractors stand up to the polders, they are good enough for most farmers. The Steiger Couger from Dakota with its 250 hp is another one of the big centre-pivot types to reach our shores.

Manufacturing firms, and also individual farmers, still come up with a hitch to couple two tractors together. By taking the front axle off the rear tractor and coupling to a suitable drawbar, one can join up, say, 75 hp tractors to give 150 hp outfit for heavy cultivation work. If that sort of unit sends chills up the spine, it is likely to be fear rather than thrills. I cannot convince myself that the idea looks right. I do not like untidy combinations, nor anything that looks as though it could be dangerous. Clutches, gears and stop levers have to be synchronised.

The thought of farmworkers doing the coupling and uncoupling is something that frightens me. Nor do I believe that the average farmer intends to go in for all this work. I know that it happens with such machines as the Standen Solobeet sugar-beet harvester, where the power unit can be taken out and put back to prime-mover form for the rest of the year. These types of machine are not my idea of the way agricultural engineering should go. Nor do I believe that the farming industry is so devoid of capital as not to be able to afford sophisticated purpose-built machines that look as though they were engineered from the beginning, not hybrids.

Massey Ferguson now go to 180 hp with the 1505. This idea of a chassis that pivots in the centre, four equal-sized driving wheels and major parts mounted independently on the chassis frame is something that has caught on in recent years. But it did not happen quickly. The centre pivot MF 1200 had been around for some time before many appeared in the fields. Farmers were sceptical, to say the least. Now, this concept has become accepted.

FINLAND'S BOGIE TRACTOR

One of the exciting moves away from the conventional has come from Finland with the Valmet 1502. It gets the name of a bogie tractor because design allows the rear axle to be lifted or lowered in a six-wheel configuration. The Valmet is not at the top as power ratings go with its 150 hp but it is claimed that with the bogie wheels on the ground, or with alternate tracks fitted, ground pressure can be as low as 5 psi or $1\frac{1}{2}$ psi respectively.

Nobody should dismiss lightly six-wheel, or even eight-wheel tractors. In the absence of a large-scale return to tracklayers, because of road-work drawbacks and because the big tractors get heavier all the time, the idea of a large amount of rubber, necessarily spread over more than four wheels, might appeal. If farmers insist on running wheels in furrows for ploughing, with the tendency to twin wheel layouts on the big centre-pivot tractors, the idea of aligning wheels as on the Valmet could look right. Evaluation work also inevitably shows the greater stability on multi-wheel configurations. Safety in farm machines, it must be remembered, is something that Scandinavians are keen on.

Farmers, although reasonably conservative, will adopt new ideas and new machinery techniques as fast as any other industry, and sooner than most. Even so, they will not be rushed into anything that looks too gimmicky, too complicated or too far removed from what has gone before. This means that any far-fetched, revolutionary design changes are unlikely to come about in relation to our prime mover.

It seems that we shall continue with a wheel at each corner, possibly with a tendency to drive four instead of two wheels and often to fit larger tyres. The operator is fast moving to the idea of a refined capsule that will isolate him from the environment.

There are very few disadvantages. It is rather like the old adage of the farmer's foot being the best manure, or the one about the stockman who is still doing his job, even if he is only just looking.

Even power steering can put the driver a little out of touch with what is happening, unless he keeps his wits intact. Tractors, unlike motor cars, are not necessarily made to be pleasant to drive. They are a means to an end. They are pushing or pulling, and it is the implement that is doing the groundwork. Putting one's feet on the soil and just looking, can sometimes, as in the stockman's case, be a good thing.

LOOKING INTO THE FUTURE

The press-button tractor is not just around the corner, though it may be only a decade away. The Japanese firm of Kubota, already selling small farm tractors and large garden tractors in this country, has a beautifully streamlined model ready to market. Still at the bottom end of the power range with only 25 hp, it has hydrostatic transmission, but remains with diesel as its fuel. Allis-Chalmers, way back in 1959, released news of an electrically-powered tractor. This fuel-cell prime mover was to have the cells arranged as required for weight distribution. Alas, with oil reserves put at a life of thirty to forty years, as against coal at 150 to 250 years, diesel (despite the oil price bonanza) rules supreme.

Some farmers say they want tractors that they can hang things on, Unimog style. British firms even try copying that principle at times, and certainly there have been cases of the forward control County being adopted as a carrier, besides being a power unit. The favourite task for engineers, exploring the possibilities of a unit that carries, is the one of lime spreading.

It has always been my view that the power unit that looked to have great chances in the future was the New Idea from America. This power bogie could be driven into a combine harvester, a forage harvester, a potato harvester, a beet harvester and all sorts of other things.

The Hovercraft gets linked to farming thought at the arrival of a wet autumn. A carrier, rather than a puller, it is of course the very

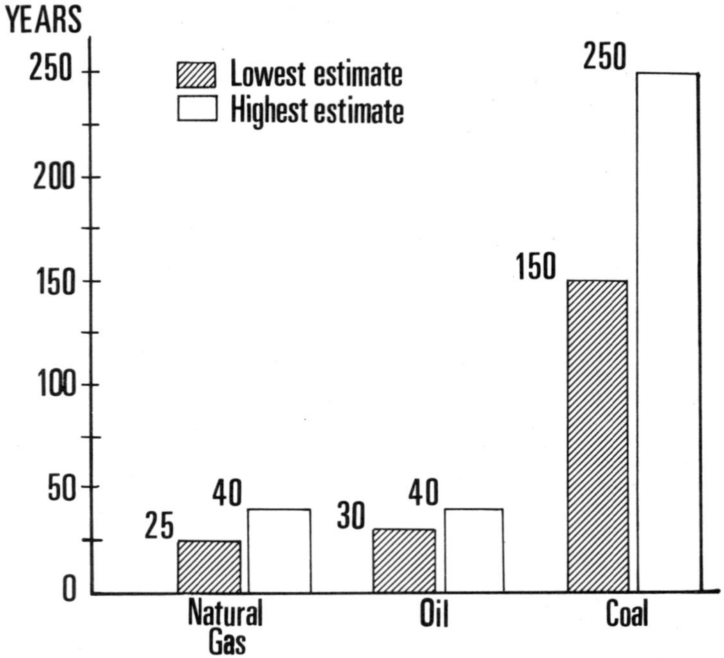

Predicted life of world energy reserves.

Source: Institute of Fuel (potential reserves future consumption).

opposite to what we already have. Whether it could both carry and pull satisfactorily is doubtful, but certainly as an alternative to aerial sowing in a bad season it has possibilities. Despite talk of turbines, rotary engines, fuel cells, a revival in steam, or a do-it-yourself methane gas unit, most realists seem to think that the reciprocating piston engine will be with us for a long time yet.

TURBO-CHARGING HERE TO STAY

Turbo-charging is here to stay as an adjunct to the in-vogue diesel. Using the engine's own waste energy, via a turbine and a compressor, we thus get a more efficient use of fuel, more power and often a quieter engine. Magnetic implement hitching and implement performance, viewed on the television screen in the operator's capsule, is hardly a certainty considering we failed to make full use of Harry Ferguson's pick-up hitch and have not advanced far with radio control back to the farm office.

The certainty with farm or road vehicles has to be more and more automation. The weather forecast to the driver's cab would need to be far more accurate than at present and is of no importance if the man in the seat could do nothing about the given situation anyway.

Always, be it noted, I am assuming the need for an operator, rather than automation on a wholesale scale. Weather forecasts to the cab, closed circuit television, electric drive and all things modern, these I regard as things of the future, rather than immediate.

CRAWLER TRACTORS

Funnily enough, although we tend to argue the merits of wheeled tractors, and especially the relative virtues of two-wheel against four-wheel drive, crawler tractor sales and numbers do not alter very much. If farmers move away from 4WD models, then contractors fill the gap.

We have had the traumatic ups and downs of such firms as Rolls Royce, Leyland, Marshall and others. Ironically, after waving farewell to the Track Marshalls, we soon had them back in the Aveling-Barford Group, which is itself a part of the Leyland conglomerate. After being introduced to our market in 1956, the Marshalls by the early 1970s were claiming a 90 per cent share of British crawler sales.

The TM75 and the TM90 are extremely popular as farming crawlers. Long overdue, crawlers now have the sophistication of three-point linkage that wheeled tractors have had for so long. Marshalls claim the first really successful crawler linkage of its particular type in the world. Draught and top links are mounted on a special swinging drawbar, pre-jointed under the centre of the tractor. Levelling cylinders can be adjusted to various positions on the draught links by the use of quick release pins, so that a choice can be made between using maximum lift capabilities for close coupled implements requiring high lift and using maximum load capabilities.

Everybody knows of the tremendous advantages of crawlers for minimum ground pressure and the maximum use of the power in the engine converted to drawbar pull. There is no longer such a big difference between tyre prices and rollers and idlers on crawlers. For wholetime ploughing and cultivation, even now (1977) I doubt whether anything compares with a crawler in efficiency. But there are the drawbacks, equally well known. There is the slower speed, the difficulty if a public road has to be negotiated, or, alternatively, the use of a low loader to transport the crawler around. Always in my mind, and harping back to the attempt by Roadless many years ago, is the thought of an all-rubber track. International, of course, besides Fiat, keep a hand on the crawler market here and back into the fray come the Caterpillars from USA. I hold the 'Cats' supreme.

FOLLOW THE LINE OF MARKET SHARE

Present-day sophistication does not vary very much from one manufacturer's range to another. The moment one finds the need for the kitchen sink to be in the driver's operating arena, then the others will quickly follow suit. Coincidence or not, at the moment of compiling this book, and bearing in mind the inevitable time lapse between writing and publication, my preference for my prime mover if one is about to mechanise a new farm is follow the line of the market share.

Ignoring all those side issues, which, one knows so well, nobody in actual fact can do, I go for the Fords by a short head, followed by Massey Ferguson, International, David Brown, then the rest of the field. Most British-made, diesel-engined farm tractors are good, some very good, but there are always some slightly better than others.

It would be far better to give operators automation for optimum output, though often with manual overriding control. Mini-computors for fuel injection, electronic ignition, automatic light dimming and so on are all possibilities for the future.

As the monsters of the prairie land grow ever larger, it is noticeable that the little dump trucks, with either diesel or petrol engines, often find their way to the farms, after initial depreciation has been taken in Industry. This is why farmers of the future will look hard at the Deutz Intrac, or the New Idea, or both systems. There are the giants at one end and the runabouts at the other. Maybe ever mindful of standardisation, something that embodies the best of both worlds will find a niche on the farm.

There is no best buy in farm tractors if all side issues—local agent, farm type, money availability, operator's IQ—are disregarded. Two factors, as I have tried to emphasise, should dominate: *standardisation and simplicity*. After that geography, relative to service and sales, is of the utmost importance. At the end of the line, in the case of most farmers, trade-in price or, if his foresight is that good, future trade-in price will play a final part.

Chapter 2

TILLAGE MACHINERY

FUNDAMENTALLY and financially the tilling of the soil could be described as creating an environment where crops are able to grow, where they can be harvested in a saleable condition, and finally sold to pay for the machinery that tilled the soil in the first place!

The rudiments of the game have changed little during our lifetime. There are more new arguments than new techniques. Farmers are not opposed to a lengthy prologue at a meeting or a discussion group, only to return home and till as father and grandfather did before them.

Just as a fortune awaits the man who can prove that what is pleasing to the palate is good for the stomach, so could a fortune await the company that can prove the folly of turning over 500 tonnes of earth per hectare to create a soil environment fit for crop development. The arguments along these lines have come to partial fruition with new techniques in minimal tillage, semi-minimal tillage or in simpler terms, direct drilling. We even get a confused feeling from the terminology.

SIGNS OF A BREAKTHROUGH

Few practical, hard-headed, self-made farmers are fanatics about any reduced tillage techniques. New methods have just about got off the ground, but no more. No-plough techniques are like automated tractors in that they are fun for reconnaissance, but only the odd farmer or two would have the courage to use such methods for total warfare. Yet the signs of a breakthrough are always there. Mainly because of the presence of paraquat and diquat in the chemical sphere, the possibility of the big breakthrough is always just below the surface.

Such a fundamental change could drastically alter machinery requirements. One would go so far as to say that, in the unlikely event of an almost overnight change, bankruptcies on the one hand and phenomenal affluence on the other could result! Imagine (if the traditionalist can even go that far) an agriculture without the plough!

Any firm geared up for direct drilling and chemical protection against weeds, diseases and crop failure would be on a winner, while firms committed to traditional tools and no diversification could be facing insolvency. Knowing farmers and farming, this is unlikely to happen, but farmers are so good at adopting new techniques, once proven, that nobody dares to discount such possibilities.

Ironically, the cautious, in-between methods of tilling, drilling and nurturing of our crops have not caught on to anything like the extent that one would have expected. Yet I remain convinced that there are middle-road tactics that would cut costs and serve us well. More recently in the UK there have been signs of younger farmers—and far-seeing farmers—looking hard at the various methods that are available. Visits to Continental shows have often given the right type of husbandman the right sort of outlook.

ROUND-THE-FIELD WORKING

The present generation of farmers know nothing but straight lines. It is surprising in fact how difficult it is to persuade a farmworker to drill a field round and round, as against up and down. At least that is the case with the good, old-fashioned types who have inherited and carefully preserve pride in workmanship. Youngsters, less precise, less conscientious and with few inherent skills, are quicker to realise that round-the-field work is often much easier and less demanding than the straight up-and-down stuff.

A piece missed, a blocked coulter on the drill, or some overlapping, is less noticeable with round-and-round working. Few people and fewer machinery designers seem to appreciate the advantages that go with round-the-field drilling if combining is going to be done in the same fashion. And it usually is. The combine dividers can often be running between rows and shedding far less grain than when travelling across rows. The farmers reluctance to leave the age-old plough technique is barely understood when we recall the present-day cost of turning over anything from 150 to 250 millimetres of top soil.

It is better understood by practical farmers looking round fields that may have previously harboured a dung hill or potato clamp or sugarbeet heap. It often happens that a small part of a field has to be missed with the plough at autumn time because of a heap of dung or a root crop still being harvested or delivered to the processor. Come spring seeding and it is common to cultivate over the whole field, including the small area that did not get the basic ploughing.

IMPORTANT OBSERVATIONS

My experience is that when one inspects such fields around June to July, especially if it has been a dry summer, these areas will show the tell-tale signs of having had no real primary cultivation. Often the crop is pinched, poor and paltry. There may be farmers who can tell the opposite story of such unploughed areas producing a better crop than the bulk. Never has this been so in my farming life. Often, in farming, with farmers being keen observers, fashions are wrought out of these small observations that would be unnoticed by the layman.

Such small but important observations can be and are responsible for trends and fashions in agriculture. If profit factors can support them, so much the better. All sorts of figures have been published from time to time, showing that minimal tillage methods give as much or more grain than traditional means. Nor is it over difficult to make a case out for cereal production via direct drilling and chemicals.

As it happens, like so much in our lives, the oil crisis for us and the boom for the Eastern world has put the price of the chemicals up, more or less in proportion with the cost of the cultivations. So if the rising cost of diesel and all subsidiary oils turned any heads towards direct drilling as a cheaper way of producing crops, then they were soon to be turned back again. In fact, chemical costs have rocketed so much in the wake of the oil prices that there could well be a case for saying that the plough is needed for soil inversion more than ever before.

CHISEL PLOUGHING

Let us face it, the backroom boys are right—we do not need the plough to make a suitable seedbed. The plough is the great original weedkiller. Secondly, it buries trash to suit the Jethro Tull system of drilling regardless of whether tine, disc or coulter is used. So far as UK conditions go, even the chisel plough, working its passes at a 30 degree angle as against the complete opposite in travel, might not have reached its present peak had it not been aided and abetted by chemicals.

Herbicides, plus the Australian type of chisel plough, have been found ideal for much autumn planting. It has proved to be cost reducing and yield usually not very different, but still the average farmer has kept to the conventional seeder. There are exceptions and again the system is often dictated by the vagaries of our climate. A succession of bad autumns in the late 1960s caused many farmers to press the fertiliser distributor into service to sow wheat and winter barley. They found that in most cases the yield did not suffer. These bad autumns did much to prove the point of timeliness. Better to throw the good seed on the furrows than not to get the crop planted at all. It also happened that winter wheat was a more profitable crop to grow than spring barley anyway.

Thus it was that a return to the old broadcasting method was very much a possibility. Some better autumns followed in the early 1970s, but the autumn and winter of 1974–1975 brought everybody down to earth with another bang. Only the really good autumn of 1975, followed by an excellent spring in 1976, saved us from another panic about soil structure.

DRILL-AND-TILL ERA?

Somewhat overdue, it would appear that we may now be heading for a drill-and-till era, following the basic cultivation of ploughing or chisel ploughing. For re-seeding grassland, and, for some reason, growing kale, direct drilling after spraying paraquat seems to be in vogue. It works, For cereals the method is doubtful and not fully proven by a succession of differing seasons. Probably the best combination is conventional autumn ploughing, or spring ploughing on some of the lightest soils and semi-minimal cultivation for the autumn cropping.

DEEPER PLOUGHING FOR BEET AND POTATOES

For sugarbeet and potatoes there seems little doubt that the deeper ploughing, with complete inversion of the soil, still gives the best results. Many successful and ultra-practical farmers, who were virtually reared behind a plough, now settle for deep cultivation rather than the mouldboard plough for autumn preparation of lands to be sown to wheat, winter barley, or winter beans. Winter beans, still a successful crop on the Essex clays, are sometimes broadcast on to the stubble, then ploughed in with the mouldboard plough but keeping fleet in depth.

Having gone for chisel ploughing and drilling, many farmers have been in difficulties with drill coulters. This is not surprising, because we have hardly progressed since the early Suffolk coulters. This, besides the extreme elements of the bad autumns of the later 1960s, has been another reason for farmers often settling for broadcasting in the autumn. Another reason why broadcasting in the autumn can be more successful than in the spring is that the notorious triumvirate of predators—pigeons, rooks and sparrows—have usually got some stubbles to feed on.

NO SUPPORT FOR DISC PLOUGH

It must be something of a paradox that British farmers have been willing to focus attention in a limited way on the chisel plough, while scorning the disc plough with utter finality. Many other countries manage to get good, almost exclusive use out of the disc plough, while we stick with a near theological dedication to the mouldboard. Because of our repudiation of the disc plough, it would appear to be a little inconsistent to flirt, albeit half-heartedly, with the chisel plough.

There are times, mainly after a hot, dry summer, when hard stubbles look to be inviting the disc plough, but rarely, if ever, does the farmer turn his affection in that direction. For one thing, we like to plough a good depth, anything from 150 mm–300 mm. That way we reckon on burying all trash and rubbish unwanted on the surface, thus making humus down below. There is a school of thought, mainly the chisel-ploughing fanatics, who will argue that such humus should be kept in the top 25–50 mm of oil, where the plant roots will be certain to utilise it.

Fashions change, but the plough, it seems, goes on forever. What changes are the tines, discs and various efforts in ironmongery that are used to change the crude furrows into a suitable tilth for planting seeds. For generations, all we knew was the blacksmith harrow, with tines straight, curved, pointed or flat. When they were worn, we simply dumped them at the local smithy, where they would be unbolted, put into the fire and hammered back into the original shape.

Unfortunately, blacksmiths have sometimes been slow to move from the fire and the anvil to the welder and the jig. As a result, farmers have had to change to more expensive factory-made equipment, not just for the original purchase, but to get service and spares. Yet many small farmers still manage to till their land with a range of heavy, medium and light harrows, made and maintained locally.

A good case can still be made for the local workman, if you are lucky to have one in your district. Perhaps the arrival of the springtine harrow in the early 1950s sounded the death knell to many blacksmiths. Yet similar implements of a tricycle layout, with ducksfoot-type tines and horse pummletree up front, could be found in stackyards. Fashions in farming come and go, as in other walks of life. Suddenly, this earlier-scorned tool was in vogue in a big way.

THE SPRING-TINE CULTIVATOR

One of the great success stories of post-war years became the spring-tine cultivator. The early ones probably failed because of the pedestrian speed of the lamented horse. This was a tool that relied on speed for its shattering effect. Tractors could now haul it at five to seven mph and, that way, it did a good job. On average, two passes got us a tilth.

Later came the crumblers at the rear, levelling boards at the front end and all sorts of variations. Provided the land was not bone hard, the spring-tine cultivator would cultivate stubbles in the autumn, it would get a seedbed in the spring, or it could be aligned to do inter-row cultivations.

Discs had been around as harrows though not as ploughs, but many farmers were almost as suspicious of the disc as a harrow, as they were of it as a plough. It would cut up the rhisomes and tap roots of grasses and docks, all would grow and the weed population would increase. There was some truth in it, of course. But mainly it seemed that UK farmers were men of tines, rather than discs. The spring tine was an overnight winner.

POWER-DRIVEN TILLAGE TOOLS

It is my guess that the arrival of the spring-tine harrow did more to put country blacksmiths out of business than anything else. Now, with escalating prices and farmers committed to the storeman's counter, it is no surprise to find some farmers sticking to their harrows and desperately trying to keep the local blacksmith in business. While most farmers are still very much committed to the spring tine, others have spurned it in favour of powered tillers. Many of us only looked askance at power-driven tillage tools because of the inevitable wear, tear and cost. Many farmers still repudiate them on these grounds, which, let it be said, are not irrelevant.

It seemed that, unlike in the harvest field, farmers were in no mood for self-propulsion with a seedbed maker. And so rotary cultivators soon went trailing behind the prime mover. Still, in 1977, the name of Howard, pioneer of rotary cultivation, dominates in this sphere. Some farmers swear by the rotary cultivator, others swear about it! Many heavy-land farmers say that it creates a pan, owing to the rotary action of the tines. If used in wet conditions, they are correct.

Working on a horizontal shaft, it is not denied that high peripheral speeds plus a wide cutting surface will take up a lot of power.

This is no job for an underpowered prime mover. A safety factor that I have found in rotary cultivation is the disengagement of the pto as one lifts at the headland. The momentum when the blades are in contact with the soil can push the tractor a little nearer the ditch than is comfortable! As in so many other instances, the hydraulically-engaged pto that can be slipped in and out so easily compared with the old tractors is a boon.

THE ROTARY CULTIVATOR

The art of conserving power with the rotary cultivator is to only do what is necessary. Rightly, protagonists of the rotary types can claim that they were really original in thinking up the one-pass implement. For catch crops or reseeding it really could be a one-pass tool from the stubble. More recently it has been used in conjunction with direct seeding, by spacing cutting tines at the desired width along the shaft. After ploughing, the rotary cultivator was found to give a seedbed in a single pass. One of the mistakes in usage is often to cut the soil up to particles that are much finer than need be. A heavy shower between cultivator and drill can result in a quagmire, and that is the sort of situation that gets the rotary a bad name. It is a tool for careful and sensible usage.

The speed of the rotor needs to be matched to the forward speed of the tractor, both for a nicely balanced cultivation and power ratio, besides preventing what is usually called trowelling. There is no need to make an over-fine seedbed on the modern types, because manufacturers now give the customer both gears and data. The degree of shattering can be controlled with suitable arrangement of the gears in the cultivator gearbox. Two wing nuts usually expose the gears, and either diagrams or the instruction book will tell how to arrange their positions for the type of seedbed required. Once the tractor driver is familiar with the gears, their colours and positions it is only a matter of minutes to change to a rougher or finer tilth.

Ordinary harrows, cultivators and spring-tine harrows are so well accepted by the farming fraternity and so common as everyday equipment as not to warrant any details of usage. One of the still common mistakes with tractor drivers—and this is not confined to cultivators— is the failure to set the top link at the proper length.

Not only is it wasteful of power to have implements crabbing, digging, or tilted, but it is obviously not good for the framework and hardly best for getting an even seedbed. It is surprising how many men will complain that a plough will not lay its furrows with the desired correlation and will tamper merrily with sideland adjustment, coulters, skims and discs, only for them, or their bosses, to discover eventually the top link far away from the recommendation.

While the rotary cultivator did not change for years, it has now been adapted for different techniques. For reduced power requirements and for making an ordinary sort of seedbed suitable for cereals, there is the Howard Rotaspike, which is what it says it is. Spikes replace the blades on the rotor. Then there is the Rotadigger, where again the claim is made that less power is needed. This is a combination implement with tines doing the primary cultivation, assisted by the rotary. The Underbuster is a similar attachment but referred to as a subsoiler, again working in conjunction with the rotary.

It is noteworthy that the Underbuster attempts to put the plough in the middle of cultivating machines, then dispense with it. With tines designed for maximum soil shattering at a good depth, plus the rotary cultivator for creating the seedbed, this sort of machine certainly embodies ways of going from stubble to seed-bed in one pass minus the plough, With the Rotadigger, it is claimed that the land can be left to over-winter, as with the traditional plough.

THE PLOUGH STILL SUPREME

Ever since Robert Ransome patented a process for making a cast-iron plough share at Norwich in 1785 and then took out a patent for applying case hardening from his new abode at Ipswich in 1803, we have been trying to usurp the whole technique. So far, with no great amount of success.

We now virtually have a situation where the new and the old methods run side by side. Nobody can say that plough makers have really rushed from one century to another since Robert Ransome! It has been a steady plod, rather like ploughing itself.

On the other hand, until recent years and despite the American book *Ploughman's Folly*, published in 1943, antagonists could not claim to have done much about usurping this established system.

There is no point in me reiterating the pros and cons of ploughing here. Suffice it for me to say that good ploughing still counts for a lot. If it is a choice between bad ploughing and no-plough techniques, the latter may be advantageous. Still, if one is going to plough, the organic matter wants to be got into the best position for rotting down. In short, it wants to be buried properly. Skimmers on ploughs are far more important than coulters or discs. The method of only having a straight or disc coulter on the rear furrow brought us up to date slightly, until one remembers that the furrow is only visible

15 A new concept in plough base design has enabled the clearances between the bases of the MF 270 Diamond reversible plough to be considerably reduced while still giving efficient soil flow and exceptionally good trash clearance and trash burying. The result is that this four-furrow size is no longer than a three-furrow conventional plough and is therefore easier for the tractor to handle.

16 The four-furrow model fitted with hydraulic reset trip legs, in the Ransomes Spaceframe 200 series of fixed ploughs.

until next time round. In other words, nice-looking, sharp-edged open furrows are less important than burying the trash.

EFFECT OF PLOUGHING SPEED

We have tended away from the old horseman method of standing the furrows up on end, as it were, for the winter. We now take wider furrows (though still often not wide enough) to accommodate the modern width of tractor tyre. One development has been to make plough mouldboards that ploughed according to speed. Go slow and there is a neat furrow, duly exposed to the hoped-for winter frosts. Motor along at a much faster speed and the furrow will shatter, almost to the state of a seedbed on light lands.

The theory is good. With the early-ploughed land that is often going in with winter corn, it is possible to drive the tractor at say, six to eight km per hour. Later on, in bad conditions, the driver may be down to three to five km per hour. Then the land will be left more suitable for over-wintering, which it is probably due for anyway. For one reason or another, all ploughmen seem to think that they have a divine right to winter frosts. Sometimes their prayers must fall on stony ground, because we do not get them!

Basically, the conception of the plough as our primary tool of cultivation does not change very much. Massey Ferguson did try squashing four furrows into a much smaller area with the diamond plough, cutting diamond-shaped furrows, but most farmers still opt for what they consider the conventional. Even the automatic re-set for the leg of the plough on stony land, as typified by the Kvernelands from Norway, did not get going at any great rate of progress. Using either springs or hydraulic cylinders, when the plough has to face up to an immovable object, the leg or beam or both are forced upwards over the obstruction. Normally, the plough body re-sets to its original position automatically.

THE ONE-WAY PLOUGH

For all the obvious reasons of wanting level land for modern machinery, not least the combine harvester, we have moved swiftly, (at least for the plough sphere) to the one-way plough. Why we use the term 'one-way' when it is a 'two-way' plough, I do not quite understand. Having got mainly onto the one-way plough in the arable east of the country, Massey Ferguson with their '85' plough, tried simple automation but without making any great impression.

When the '85' was lifted on the tractor linkage, it automatically

17 The Ransomes bar coupler cross shaft is attached to the tractor lower links so that the tractor is reversed to the cultivator and the linkage raised to engage guide plates in the front of the implement.

tripped. Not surprisingly, two factors killed the theory. One was that a plough often had to be lifted during work in bad going. This meant a double lift to get it back to facing the correct way or, alternatively, the driver went merrily on his way with some furrows going the wrong way until he next looked rearwards! Secondly, my own assessment was that this method was positively dangerous, especially in the farmyard. At tractor-filling time and plough adjustment or repair moments, the sudden swing over when the tractor linkage was lifted could sometimes surprise those familiar with it and frighten any strange bystander.

So we now go for the manual turnover on the smaller ploughs and the hydraulic turnover on the larger, more expensive types. Assuming we all want comfort cabs at low decibel ratings (and that will be the law anyway), obviously all such controls on the plough, as well as other implements, will have to be remotely controlled from the seat. Offset frames are available for those who need to get up to the hedge. Some such frames are marketed as suitable for ploughs, ditchers and so on.

MOVE TO SEMI-MOUNTED PLOUGHS

As ploughs get larger and heavier, it is necessary to go from mounted to semi-mounted. Taking the Lemken as an example, their four- to six-furrow semi-mounted ploughs have all controls, including turnover mechanism, hydraulically controlled from the seat. Krone, the German firm with a base in Suffolk, have reversibles built on the sectional principle. The basis is a single furrow unit which can be built up to a four-furrow plough. Many of the ploughs marketed by the large British firms are manufactured by Huard, the French firm.

This mass-marketing technique by firms national, international and multinational is very much the order of the day. It fits in with modern thinking but makes a mockery of the different marques. The Huard ploughs are a typical example, but we would like to think that one of the outstanding contributions to that end is by the British firm of Bomford & Evershed. Just as Massey Ferguson, International, John Deere and Salopian-Kenneth Hudson use the Huard ploughs, so Massey Ferguson, International, Leyland all market Bomford & Evershed.

I rate the success of the Bomford & Evershed tillage equipment as one of our important developments in the post-war years. They market the implements in their own colours, just as the trailer manufacturers do, while colouring some models that are contracted to the big boys in the prime mover market. Just as the plough makers give the customer shims to alter his furrow width, and Massey Ferguson put their shims on the side of the mouldboards on the Diamond plough to reduce length, so Bomford & Evershed go for various tine configurations over their range.

Single, conventional leaf springs gets supplemented by triple-leaf, square section and the massive chisel plough tines. Nor is it difficult to arrange the tines as wanted on the frame for a specific purpose, such as row-crop work. Just as farmers refer to crawler tractors as 'Caterpillars' from a sub-conscious respect for that breed, so do many still call their spring-tine cultivators 'Triple K' because the Triple K brought in by Kongskilde to their base in Norfolk soon became as common as a Ransomes plough.

Since then Kongskilde have opted for other tine types that will accomplish heavier work. Notable is the heavy duty two-piece tine where a breakage is likely to result in buying only half. This cannot fail to appeal! There is just the point that it is not always a good idea to go chisel ploughing with a tool that was not designed that way. Heavy duty tines are good stubble cultivators for the autumn, but will often go deeper than required in the spring on land already ploughed. Even discs that do a good job for autumn seeding will so often penetrate far too deep in the spring, especially if the desired amount of frost infiltration has taken place.

DISC HARROWS

Disc harrows are another tool that some farmers go a farming life without and others do practically all their seedbed preparations with. It is not too difficult to go through a lifetime in farming minus a set of discs. Apart from some sealed self-aligning ball bearings, the Bamford-Kvernelands that will fold for transport, and the Pettit that gets called a reversible disc plough and cultivator and has had some minor changes, the disc (like the plough) has remained much the same. This could be a tribute. Maybe the plough and the disc harrow were so perfect at their inception that no worthwhile modification has been necessary.

Serrated edge discs supposedly cut through trash better, and it is common to have them on the front gang with plain discs on the rear. I have noticed no great advantage or disadvantage in the various combinations. Much the same remarks apply to farm rolls. We have gone from one roll to triple gang rolls but have done little to assist the man who has the job of coupling them together. If roll makers had to couple and uncouple on a dirt base, as against a concrete base, it could be different.

RIBBED-TYPE AND FLAT ROLLS

Taking the Hunt rolls as an example of the ribbed type and the Kidd as an example of flat rolls, for many conditions it seems that arguments are superfluous. On the lighter lands we reckon that

18 With the Ransomes C92 cultivators a basic frame can be extended as the farmer increases the power of his tractors; there are five options in bolt-on tines. The transport kit seen here enables the wide versions to be towed lengthways.

19 The extra-strong two-piece tines of the Vibro Flex cultivator from Kongskilde.

20 The Lilliston rolling cultivator for inter-row work brought from America by Opico (UK) Ltd.

21 The Lely Roterra which revolutionised seedbed preparation with its ability to get a seedbed in one pass.

ribs are necessary because the land is not so prone to blowing that way. It could be just as easy to say that the flat rolls cause a crust or pan to form on heavy land, that prevents the necessary penetration of water. The makers of this simple piece of equipment have not kept pace, despite the simplicity. The Clevis hitch could have been replaced by the pick-up ring long ago. Many farmers immediately fit their own ring-type drawbar. I would have thought that the use of hydraulic power, either to raise the outrider rolls (or 'pups' as they get termed at farm level) or to park them in tandem for transport, was long overdue. Again, it could be that rolls are so simple in their conception and application that no improvement or modification is called for. But manufacturers could listen to the language of a farmworker at coupling-up time!

USE OF HYDRAULIC POWER

A good example of harrow folding by means of hydraulic power comes with the Parminter, while the Cooper Octogang disc harrow also folds by hydraulic ram. It is imperative, as implements get wider, that this sort of sophistication becomes a common trend in the province of cultivation machinery. There is no doubt that the cost of labour and fuel has had a marked effect on farmers' thinking over the past year or two. There have been attempts at reform in the shape of the Turbotiller and the Sampo rotary harrow, besides the Lilliston rolling cultivator, without any real trend being created.

Not so with the powered cultivators that came into their own when nobody answered our prayers for frost and snow! The Benedict and the Vicon power harrows were a revelation in their early days. I am old enough to remember Wilders Cultharrow, and, of course, the Pitchpole harrow, which in its day was outstanding.

The Pitchpole had straight tines that did a superb job on stringing out couch grass roots, while the short-lived Cultharrow, that the great Ferguson even liked the look of, embodied the reciprocating principles of the types we have now come to accept. Perhaps we put the Gyrotiller of the 1930s in too deep. With contra-rotating tines we do the same sort of stirring action, but to drilling depth only, not sub-soiling depth.

The Benedict Soilmaster, the Vicon four-bar harrow, then the cheaper two-bar model, got reciprocation in tillage tines off the ground. Or rather, they got the principle into the ground. Vicon, in the heart of arable farming at Ipswich, jumped into the power harrow limelight in 1965 after one of those winters when God refused the ploughman's request for sufficient frost to put his furrows right.

Chance's chisel plough with its Australian connotation, the rotary tiller from Finland, sub-soilers of various types, the rolling harrow, Lister's excellent chain harrows, the Cousins levelling harrow, rotary tillers, spring tines, straight tines and angled tines have all come in with varying degrees of success. But when the reciprocating harrow got supplemented by Lely's contra-rotating Roterra, the whole concept of power harrowing seemed to take on a new dimension.

THE ROTERRA SEEDBED MAKER

The Roterra is the cultivating discovery of our time. It is linkage mounted pto driven, with a series of vertical rotors each fitted with two shaped tines which stir the soil without bringing up the unwanted stale or wet soil from below. Like the rotary cultivators, the Lely has a gearbox, so that two rotor speeds are possible and there is a levelling bar up front and a crumbler behind. It is not very satisfactory to try using the machine minus the crumbler, which is required for depth control despite the three-point mounting on the tractor. It is better to reduce the eight bars to four by removing alternative bars from the crumbler if the seedbed is too fine. Like the power harrows and rotary harrows, the Roterra does a job too good for some circumstances. Also like the others, it is a good practice in our climate not to cultivate more than can be seeded in a day. In my view, the Roterra is the real once-over seedbed maker. Even for anything as delicate and precise as sugarbeet, the Roterra can make the perfect bed in a single pass. Expensive to buy and not particularly cheap to run, the implement does a superb job on virtually any type of terrain. After winter frosts and no tilth problems, a good case can be made out for simple draught implements, as against the heavy expenditure on power harrows, rotary harrows, or the contra-rotating harrow. But taken over the years, especially on heavy, recalcitrant land, the farmer with a powered harrow will always get his seedbed.

I am taking the Roterra as the perfect example of a seedbed maker, but, as with the power harrow and the rotary harrows, it is not a monopoly market. The La Ronde, from France, is similar to the Roterra. The Tornado from Alpha Accord is another, and many will remember the Rotaspa spading machine that Vicon tried over here. Its rotary digging action with revolving spades, which turn sideways to invert the land, took us as near to the gardener digging his plot as we are ever likely to get. Alas for Vicon, the remunerative success of their power harrow was not to be repeated. Nor can I see the spading machine coming now.

Far too late now some firms are making the type of combination tools that many of us envisaged long ago. Some straight, heavy tines up front, followed by some lighter spring tines, with crumbler at the rear. There are various variations of that sort of theme. Ahead of the power harrow and the Roterra they would have been in with a good chance. To a limited extent they will catch on now, but, as I have indicated, the tillage tool manufacturers must consider this idea as one that got away. It is interesting to note that the NIAE has always favoured a rotary digger, or tine and rotary combination. Indeed the Howard Rotadigger, mentioned earlier, was developed along the course chartered at Silsoe.

THE SUBSOILER

I have dwelt on what I consider the machines of tomorrow, rather than those of yesteryear, and religiously avoided a lengthy discourse on ploughing. There is little to be said about straightforward tillage tools that have been with us for generations. Another weapon in the battle to preserve soil structure, even to supersede the plough, is the subsoiler. These now come with one, two, three or more legs, some rigid and straight, others with wings, plus a few with the pto drive vibrating the working part for more shattering and, we get told, less draught.

Draught reductions of up to 80 per cent may be claimed for some fore and aft oscillation of deep cultivators, such as the subsoiler. Yet the total power input of draught added to that used at the pto will be much greater. Farmers often buy machines knowing full well that the power requirement is more, but they may have the power available and simply aim at getting a job done quicker or more efficiently.

Several years ago I mole-drained 40 acres that were a quagmire after sugarbeet. The operation could just as easily be termed subsoiling, because in effect that is what it was. We just wanted to rid ourselves of the surface water to be able to plough. I knew of only a few drains in the field, which started heavy at the top end and finished light at the low end. The effect in the following barley crop was so marked that I called in my Ministry drainage officer to witness the spectacle. Strips about 600 mm wide were to be seen growing a crop often as much as 300 mm higher than that between the moles. It was most noticeable on the light land.

Thus one can only applaud the belated attempt to draw the farmer's attention to the pan below our feet. The subsoiler is the correct tool for such jobs, but in many cases the subsoiler bottom

22 The Lely Combi 300 enables one man to prepare a seedbed straight from ploughed land and drill at the same time. The Combi has 17 Suffolk-type coulters if fitted to the Roterra 300 (3 metres wide).

23 The Vicon four-bar harrow will make a seedbed in one pass under most conditions.

24 The Bomford & Evershed Superflow, behind this Track-Marshall 753 is typical of the range of tillage implements in the semi-minimal sphere. The Superflow is often used in place of the mouldboard plough, especially for winter sowing.

25 The Canadian 'Packer' now available in the UK. This model is fitted with weight trays which can be filled with any suitable material including soil or clods from the field being tilled.

is interchangeable with the mole drainer. As I have indicated, they are both versatile tools that allow either cultivating tines or just one or two subsoiling legs to be fixed on a common frame, besides all the other interchangeability. Wings on the subsoiler bottom are little more than welded-on hunks of metal, and the idea of using the pto for vibration gets tried from time to time.

TRAILED AND MOUNTED TYPES

There are subsoilers trailed and subsoilers mounted. The technicalities about shattering effect and power consumption are far less relevant than the principle of usage. The subsoiler is cheap, simple and effective. Farms are mainly comparatively small, one-boss businesses. The decision to try out a new tillage technique, or to revive an old one, does not cost a fortune and can usually be done by a contractor if necessary. With subsoiling, quite regardless of the finer points of the techniques argued about, there just has to be one predominant thought in the mind: even a one leg unit needs a lot of power to go to a sensible depth of, say, 500 mm.

Apart from the lightest land in near-perfect conditions, I reckon that any farmer with less than 75 hp under the bonnet of his tractor is better to leave subsoiling to the contractor. Ideally, especially with mole draining, this is a job for a crawler or at least a four-wheel-drive tractor. A mole drainer, with the expander fixed behind the mole, is very much the crawler tractor's prerogative. The best mole draining was done by the old steam engines, where the drainer was pulled uphill only.

I do not consider field drainage to be within the compass of this book, because this is mainly a specialised job done almost exclusively by contractors. But, since the inception of co-partnership and sharing arrangements with machinery, I have been surprised that heavy-land farmers have not gone into jointly-owned drainage equipment. Perhaps that alone explains that it is a specialised sphere. There is also the question of drainage grants and approval of schemes by the Ministry of Agriculture which also tend to put this operation into specialist hands.

FARMING WITHOUT THE PLOUGH

One can summarise tillage by saying that a beginner no longer needs to own a plough. It is feasible and possible for a young man to start farming today and boast that he never owned a plough, just as I boast that I never owned a horse! On the other hand, I doubt whether there is a lot of financial advantage in not owning a plough. It is still one of the world's best destroyers of weeds—used properly, that is. If the alternative is deep cultivating, or, if you like, chisel ploughing, then there may be some advantage, but not very much. None perhaps, if extra passes have to be made.

But there is one advantage bearing in mind present-day machinery costs: a beginner can now do his tillage with far fewer machines. It would be quite possible for an arable farmer to manage with a cultivator, powered harrow and a spinner for both seed and fertiliser. That would not necessarily make one the best farmer in the district, or the tidiest, but it could be the means to a satisfactory end—that of making financial ends meet.

Spring-tine cultivators are much of a likeness now, but one has to retain that sneaking regard for the Triple K. But whether spring-tine cultivators, chisel ploughs, or any tined cultivator, there can barely be a failure in the Bomford & Evershed range, in addition to those marketed by the tractor makers and made by Bomford & Evershed. Two chisel ploughs I have regard for are the Scimitar from Mechanaid and the Vicon. In rotaries one cannot go wrong with the name of Howard, but another machine that has impressed me is the Kronevator from Bernard Krone (UK) Ltd. If I needed a power harrow I would decide on the Vicon two-bar or four-bar according to the land that I farmed. Discs vary little but the big tractors seem to have brought the best out of the Pettit and the Cooper. But, in my judgment and until something surpasses it, the Lely Roterra is the outstanding seedbed maker. Despite the years between Robert Ransome and 1977, I find it hard to find ploughs with the all-round ability of this particular range. If I wanted a big plough for one of the giant tractors, I would choose the Dowdeswell. While rating the Dowdeswell highly, we have to be fair. Not for the first time, the groundwork comes from another manufacturer. In this case it is mainly Ransomes mouldboards and ancilliary wearing parts, but some go out with Lemken bodies fitted. Lemken themselves make a pretty useful range of ploughs and other tillage implements. The Huards, from France, marketed mainly in the colours of some of the large manufacturing conglomerates do a satisfactory job at a fair price.

Chapter 3

DRILLS, SPINNERS AND PLANTERS

WHEN THE International and the Massey-Harris combined seed and fertiliser drills came into this country, we had a lot of phosphate-deficient land. This had been caused by the wartime ploughing programme that had brought so much old pasture under the plough.

The early drills were not so well designed as today's seeders which enable the man on the tractor seat to see whether the seed is running. So we had and still have, for that matter, the theory that a second man is worth his time to see that there are no blocked coulters. Many drills had their mechanism geared in halves, so that it was possible to have faults in one half of the implement. I once had a field seeded with half the drill putting out seed and fertiliser, while the other half was drilling fertiliser only. There was a man on the back! To be scrupulously fair, unless the man on the rear of the combined drill that had seed and fertiliser going down the same spout had eyes as good as a hawk and was supremely conscientious, that sort of thing could easily happen. Nowadays most drills have seed and fertiliser going down separate spouts, though I am not convined that the additional expense involved has produced anything worth while.

A man on the footboard now has no excuse for not seeing whether all is well. But a far better check, with the drills having both materials travelling down the same tube, is to drop the coulters into work at each filling. Then walk along the back checking that both seed and fertiliser are seen at the bottom of each tube.

TWO SPOUTS OR ONE?

I look askance at the two-spout advance in combined drills, not because it is not a worthwhile feature, but because unless somebody is foolish enough to combine-drill peas or beans, there have been few cereal crops damaged through the seed and fertiliser being in close proximity. When it did happen in one particular season and people rushed to point to the danger of dry conditions and the seed not germinating quickly, others considered it to be too much urea in the fertiliser. If that was the case, then the combine drill was exonerated.

We have now advanced far beyond that great Jethro Tull strategy of 1701 that gave us the seeding in rows, reduced seeding rates and the ability to clean the crop afterwards. Sprays have, thank goodness, done away with horse-hoeing, tractor-hoeing and hand-hoeing. The drill goes on, sometimes it seems by the skin of its teeth, but at other times it looks as firmly entrenched as ever. Now, should we so wish, it is possible to put the fertiliser down one side, behind, or below. Theories and reports of better yields for this method or that have come along over the years. But, in my view, the combine drill did as much as anything to improve yields in the immediate post-war years.

Farmers have gained far more than they have lost by the rare case of germination damage owing to the close proximity of seed and fertiliser in the soil.

COMBINE DRILLING

As the phosphate content of our soils is now back to par, there is now little to be gained from combine drilling in most cases. Rightly, one feels, farmers are going for the straight-forward and simple methods. Failure to use proper maintenance procedures has made many farmers move away from the combine drill, criticising it bitterly when they should often be castigating themselves. Others tend to do the two operations separately because they know that there is now little to be gained yieldwise.

Many farmers plainly and simply find that they can do their drilling and fertilising quicker and thus more economically than the slow and rather cumbersome method of combining the operations. One of the advances in farming techniques has been the speed with which the ground can be covered with either seed or fertiliser. We do not have to have the two commodities on one trailer or lorry. Drilling does not have to stop with the first drop of rain, as is the case with fertilising. Liquid fertilisers have played their part because not many farmers contemplate combine drilling with liquids. In fact, it can and is done by Chafers, but one does not see any great rush into that technique.

The speed over the ground argument has even caused many farmers to re-think all the post-Jethro Tull methods. Drilling in 180 mm rows, whether combined drilling or not, had become almost standardised. Then came the 'grain only' drills, mainly from overseas, that were usually drilling rows of 100 to 125 mm. By this time fertiliser distributors had gone mainly, though not entirely, to the spinner type, as compared with the full-width type spreader.

FORCE-FEED DRILLS

To deal with the seed side first, it has to be said that, however successful in its day, the good old cup feed of the Smyth type of drill has had to give way to force-feed drills. When tractors began to replace horses, and the extraordinary steerage framework, plus chains and other impedimenta, were replaced with a drawbar, the tractor's speed soon had half the grain flying over the top of the cups. For sugarbeet and such crops where a sedate pace was needed for drilling accuracy, the Peasenhall drill and other similar breeds could remain. For cereals, beans, peas and similar crops, the force-feed gradually took over.

Whether external force-feed, internal force-feed, fluted roller, studded roller, it matters little. What matters more could be whether we drill at all. But we will leave that for the moment. There has always been some slight challenge to the force-feed drill from pneumatic or centrifugal action drills. It has remained slight.

The latter both have the advantages that the spinner has over the full-width spreader, in that effective working width can be far greater than the width of the seed container.

It is not quite the same in that there has to be some mechanical structure where the spinner has nothing, but at least it can be of light-weight construction and easily folded for transport. On spreaders we want that definite advantage, on seeders we do not. Farmers are funny people! Drill manufacturers and accessory makers soon cashed in on that! The farmer who has to negotiate gateways or narrow roads, or for that matter, motorways, can spend more money on a drill transporter. With a bit of ingenuity in the workshop, it is not beyond the talents of the farm mechanic to construct such a contrivance. With a bit more dexterity, and if both uses are required on the farm, it might be possible to make a carriage suitable for the drill and the combine header as well.

Apart from some risk of cracking large beans, the force-feed drill is doing its job satisfactorily. Cracking winter beans or, if you like the phrase better, horse beans is not conductive to good germination. It is a good idea to check on what is the largest size of grain a given drill will deliver without risk of seed damage.

The Scandinavian rotary action that uses centrifugal force to deliver the seed to tubes fixed in a ring round the cylindrical seed container, exemplified in the Jones and persevered with by Bamfords for a long time, has been widely accepted. The same can be said of the pneumatic models, using the Alpha-Accord and the Bamlett-Tive as examples. This type meter the seed into an air stream for delivery to the coulters via flexible tubing of one kind or another. Recently Vicon, famous for their pendulum action spinner (or 'the waggling spout' type, as most farmworkers prefer to call it) have adopted the pneumatic type in conjunction with Alpha-Accord.

In the old days, one drill on the farm sufficed. Now we tend to segregate into two camps. One drill for cereals and pulses, the other a precision drill for sugarbeet and all such crops. Maize, of course, is a crop that has come into the reckoning these days and gets put into precision seeder bracket. Different coulters have come, but few have gone.

The Suffolk coulter that was once fairly universal has partly gone, but still many farmers prefer it, despite its bad record for penetration. Provided the seedbed has little or no trash, the Suffolk type will drill evenly and well, but for most conditions the disc coulter would appear to be the favourite. The disc does not, I submit, drill as evenly as many would like to believe and certainly its ability to cut through trash will depend on the firmness of the soil. On the firm land—and surely a seedbed rarely is—the discs cut through trash well.

POPULARITY OF TINE DRILLS

The double-disc coulter, as against the single-disc which could be termed the conventional, is used mainly in direct seeding drills, as on the Bettinson. Tine drills, no longer the prerogative of Australia or the Western Hemisphere, have come to the fore in recent years. From the first, the spring tine cultivator had seeding possibilities and it was no surprise to me when the Carier, from Essex, appeared on the scene. International use tines on their model that is designed for direct drilling, though such tines naturally have to be much more substantial.

Tine drills are very uneven with their seeding, but whether that is any great disadvantage is someting that will eventually be decided. There are two schools of thought. At one end of the scale many farmers have had the courage of their convictions and given up drilling in favour of broadcasting. Like giving up the plough, there is something sanctimonious about giving up the drill. Which means that not many have done it.

At the other end of the new thinking are those who believe we have to follow in the wake of the trend set by sugarbeet growers. In other

26 The Ransomes-Nordsten 3-metre cultivator drill.

27 The Stanhay Precision cereal drill makes available to cereal growers any advantages obtained from precise spacing and programmed depth that have been available to growers of roots and maize for some years.

28 The smaller version of the Alpha-Accord drill.

29 Bettinson 3D Direct seeding drill putting kale direct into stubble. Unique features include rubber buffer springs, synchronised hydraulic depth control, triple discs and twin neoprene roller seed metering.

words, we reverse the historical evidence by putting all drills into the precision category. Before Jethro Tull, they were all in the broadcast section. There is just the point that no metallic substance came into the design in those days. Only the muscle structure of the human arms!

It would be economy par excellence to go back to all broadcasting for cereals and pulses. There is no guarantee that this will not be the case. Or we could go for direct seeding using the double-disc coulter of the Bettinson-type. For stony and light soils one supposes that we could get away with tines of various types. Other manufacturers, mainly Stanhay from Kent, have decided that we could get better yields from our cereals if we treat them like maize or sugarbeet and drill with a delicate precision.

DIRECT DRILLING

The Moore Uni-Drill, that won the Burke Trophy at the 1976 Royal Show, is an 18-row close-spaced drill that lends itself more to direct drilling than the conventional. It has not got double-discs coulters but its single discs will give a penetration variation that makes it suitable for cereals, brassicas and grass seeds. It has a simple wheel drive that folds for transport and, unlike the old disc drills, it aligns itself with the latest in sealed-bearing fashion by having no grease nipples for that unwanted daily chore.

In an attempt to be futuristic, let us look at the precision drill for cereals. As I have indicated, there are distinct signs of a majority swing to one extreme or the other. Much will depend on the price of cereals, which is reflected in the price of seed corn. The higher the price of seed, the more justification for the precision drill which is economical on seed sown.

OPTIMUM SPACING FOR HIGHER YIELDS

Since 1971, extensive trials have been carried out, which show that yield increases can be obtained by having an optimum plant population. Often seed rates can be reduced without loss of yield. In general optimum spacing for top yield would appear to some from a seed spacing of 100 mm by 25 mm. In other words, a planted seed population of $3\frac{3}{4}$ million per hectare. Farmers are not expected to actually count them! The trials, carried out by Stanhay and ADAS, extended from South England to Aberdeen in the North.

The trials carried out between 1971 and 1975 show an average yield increase of $12\frac{1}{2}$ per cent over the same seed rate drilled by conventional methods. It could be that the drought conditions of 1976 have confused the data slightly, in that a 35 mm spacing in the row, as against 25 mm, have given the better results. Be that as it may, seed spacing is not the only factor. Depth of planting could be of almost equal importance. Seeds planted beyond the elongation distance of the coleoptile, use up too much energy, or are under too much stress before emergence. The tillering point is consistent in being about 25 mm below the ground surface, irrespective of seed depth.

Fisons have also carried out trials which show that with precision drilling, tiller counts and ear counts are often greater than with orthodox drilling methods, even where weight of seed applied is less. This can be seen from the table below.

Precision Drilling—Abbots Ripton 1975
(Variety—Maris Mink)

	Grain yield tonnes/ha (cwt/acre)		*Ear count M^2 (sq yd)*	
N rate kg/ha (units/acre)	Conventional drill	Precision drill	Conventional	Precision
0 (0)	4·96 (39·5)	4·19 (33·4)	765 (640)	733 (613)
50 (40)	5·52 (44·0)	5·59 (44·5)	792 (662)	894 (747)
100 (80)	5·69 (45.3)	6·41 (51·1)	820 (686)	1031 (862)
150 (120)	6·06 (48·3)	6·41 (51·1)	902 (754)	1030 (861)

This experiment also showed that increasing the N rate also increased ear numbers, and the yield advantages of precision drilling on this site were greatest at the higher nitrogen rates. Further work in 1976 examined the possible interaction between nitrogen and drilling method.

The reasons for such yield advantages from the precision drilling of barley seem to be associated with a reduction in mutual shading, which allows each plant to compete on equal terms for light and makes for better tiller survival. Equally, the roots can exploit the soil to better effect as there is equal competition for water and nutrients. The uniform depth of drilling achieved may also assist establishment.

PLANTING DEPTH CRITICAL

With the semi-dwarf varieties of wheat that are now on the market, planting depth is even more critical. In practice, 25 mm has been found to be the optimum depth. An important point here, and one which most sugarbeet growers have long since inculcated, is that such data has to be related to each season. With beet, they can be virtually laid on top of the soil and provided plenty of rain follows, there will be sufficient germinating for a crop. Lay them on top and get weeks of fine weather and you may have a crop failure, or, at least, a re-drilling job.

Whether beet or cereals, the experience of most farmers is that the seed has to be drilled into moisture. Thus depth of drilling with any sort of precision seeding will vary slightly from year to year. As I mentioned earlier with regard to force-feed drills and pulses, an important fact of drilling, which probably gave rise to the early cup feed drills, is that no seed damage takes place. Damage is liable to destroy the embryo or endosperm and prevent growth.

Going back, for a moment, to the 1930s, there will be many farmers who remember that it was customary to sell all the best grain that came off the drum spouts, while retaining the 'seconds', and 'thirds' for farm use. The farm use for 'seconds', believe it or not, was often for seed. Economically, this helped in those dire and troubled farming times, because one got the maximum return from the market (such as it was), while the grain that was virtually unsaleable, or only fit for chicken feed or grinding, provided the crop the following year.

Let it be stated, that just as the innards of brown eggs and white eggs taste alike, so next season's quality of wheat, barley or oats will bear no allegiance to the colour, size or quality of the seeds sown. Seed rates of $1\frac{3}{4}$ million per hectare, rather than the given optimum of $3\frac{3}{4}$ million, have often given results equal to conventional drilling at normally accepted rates. It is a fallacy to think that a large seed will produce a large kernel of grain next year, or to relate the weight of seed planted per hectare to the weight harvested later.

UNIFORMITY OF SEED IMPORTANT

What it amounts to, is that, as in the case of sugarbeet seed, if the precision planting of cereals does become the accepted form, then there will have to be some precise grading of seeds. It will be less important that they are large, medium or small, than the fact of them being uniform, for mechanical reasons. Let us at least give our forefathers the credit of being far more perceptive and far more economical than many of today's farmers, by planting the 'seconds'. In fact, assuming the old seed rate of what we used to call three bushels per acre, one does not have to be genius to see that the 'seconds', always assuming equal germination, would give more original tillers.

Even though a proportion would get crowded out on the way to maturity.

Simply put, we now try to avoid the cost of those emerging plants that will get crowded out in the competition for air, light and water. I can understand the modern term of tramlines in cereals somewhat better than the blueprint system. Again brought to basic terms, if everything from precision seeding to fertilisation, timeliness and the weather is planned ahead in blueprint fashion, then the yield obtained (in any crop) will be well above the national average and even above what we accept as the best.

While I am trying to relate machinery and machinery usage to the realities of making a living from farming, I have no intention of diversifying my subject into the realms of, say, the Laloux method (in Belgium and France) of producing the absolute maximum in cereal yields, however creditable. But beginners should always be warned that in farming you strive for the earth first, the moon second! Timeliness is one of the great factors of arable farming, on any sort of soil.

SNAGS WITH PRECISION DRILLING

It is here that the precision seeders have sometimes fallen down. While we are striving for perfection, the bread and cheese that pays the housekeeping and keeps our offspring in clothes can be going down the drain. In drilling sugarbeet, we find it difficult to break the 6.5 km/hr barrier, and 4 km/hour, even now after several years' experience, would appear to be nearer the optimum. Fortunately, there may be other snags with precision drilling of cereals, but speeds have been obtained of up to 11 to 13 km/hour. Nobody should underrate the importance of that.

Still dwelling on what many readers may think to be a hypothetical matter, but offering no apologies because one has to try guessing the road ahead, however foggy, I want to state the mechanical advantages of the precision drill. There will be no need to reiterate them in relation to root drills, where such facts are generally known. One of the big inaccuracies of our seed drills is the distance from feed mechanism to ground. To me, this has always been excessive.

A short drop from metering mechanism to ground must make for uniformity. By reverse discharge (opposite to direction of travel), roll and bounce is minimised. Even with sugarbeet, the factor of rolling and bouncing on impact with the ground is known to play a not inconsiderable part in drill performance. Weight of seed over the selector mechanism is not affected by the amount of seed in the hopper.

The individual units, as with the now established root drills, will follow ground contours for accurate depth and even placement of seed.

SEE THAT SEED IS WELL COVERED

Whether cereals or roots, it cannot be over-emphasised that importance must always be attached to covering a seed that is so fleetly planted and that the soil above needs to be firmly pressed onto the seed. Usually this is done by the seeder itself being fitted with rollers. Unfortunately, in the case of roots, it is these very necessary rollers that often delay the drilling. It takes a very few spots of rain to stop the operation. If we were more precise with our weather forecasting, I could see a case for rollers that are easily lifted out of action, but present design tends to exclude that. If we knew that a fine spell was following immediately behind the wet interlude, then either rib or flat roll could firm the soil on top of the seeds. To roll or not to roll and to risk 'capping' or creating a crust is something that is half judgment and half luck. Or it could be directly related to the individual farmer's prowess as a weather forecaster. Come to think of it, a lot of farming depends on just that!

As I see it, one of the great disadvantages of the precision drill for cereals is the matter of filling. With present design, there appears no alternative to individual hoppers, to retain the advantage of units that follow the ground contours. I suppose that it would not be impossible to have a large primary hopper, with the weight factor bearing on axle or frame, feeding the unit hoppers as secondary hoppers. But let us be fair on design so far. Precision seeding is likely to show a one-third saving on seed cost, and there are now models that hold enough seed for 3.25 to 4 hectares before the next refill.

DISADVANTAGE OF INDIVIDUAL HOPPERS

A bigger disadvantage is probably the very necessary fact of the individual units and thus the individual hoppers. It is a somewhat more inconvenient job, and obviously more time-consuming unless special pre-packs to suit hopper capacity can be arranged, to fill a lot of small individual hoppers than one large one. This cannot be denied, but, given time, it will be overcome. From time to time we get bright ideas about seeding. One of the most promising ones is to plant phosphorus-coated barley early. It is not beyond the realms of possibility to coat the seed with the nutrients required. Drilling in a mulch has long since been tried.

Ideally, we want to be able to drill the corn coated with all the fertiliser, herbicides, pesticides and insecticides, to say nothing of trace elements, then shut the gate and disappear to the South of France or somewhere similar, returning at harvest. For the time present, one of the simple and effective methods is to produce a secondhand concrete mixer, put in the required amount of seed and fertiliser, despatching the mixture to the spreader for an all-in broadcasting operation.

It is not advisable to mix more than can be spread in a day because if rain hits the farm, the farmer may be in a bit of trouble with his mix. In the past, I have mixed seed, fertiliser and small seeds for undersowing. Alternatively, small seed boxes are available for many drills.

DRILL-AND-TILL IMPLEMENT

These haphazard seeding methods are the very opposite to precision drilling, but, as I have stated, this seeding job could go to either end of the scale—precision with expensive drills (and they have to be expensive) or roughshod methods based on foolproof economics. I have always been a disciple of a method that is slow to catch. This is another technique that is the antithesis of expensive precision. I refer to the drill-and-till implement that just needs the primary cultivation ahead of it.

Whether ploughing or deep cultivating, after the original pass or passes, the idea is to go onto the rough land and, in one pass, to bring about a fine, planted seedbed. There is a slight trend to the bridge method of coupling a drill to a once-over type of cultivator. That way one has something like the power harrow or reciprocating harrow on the tractor's three-point linkage, pto driven, then an up-and-over superstructure of metal to which is attached the seeder unit. Like the tractor and trailer, it is not a very elegant affair. It looks a little clumsy and the towing frame, or bridge, looks superfluous.

Advantages are that the cultivator part and the drill can both still be used either separately or jointly. I must say that this is the only way that the system can be got going in the UK because farmers look askance at the more purpose-built machines. Italy, in fact, has the Cantone, which is a complete drill, till and fertiliser machine that can take the farmer from stubble to his next crop planted without resorting to what he may think the more risky direct seeding. It is available both as a trailed and a self-propelled machine.

Farmers have not yet progressed to that advanced thinking, but nobody should rule out such machines in self-propelled form should our present prime mover, the tractor, ever lose its grip. We touch the

30 For wide versions of modern drills, manufacturers sometimes offer a transporter as an extra.

31 The 6.7 m width of the new 30-row model of the MF 30 drill showing the massive tapered box section beam frame and the individual lever control on each seed unit.

32 Stanhay 4-row S870 precision seeder, used mainly for sugarbeet, is here fitted with a granule applicator.

33 The Howard Rotacaster seen working as part of a combination machine where cultivating, seeding and spraying can all be done in one pass.

edge of till-and-drill with the Carrier drill, plus others that are similar. At Continental shows it is common to see the Nodet drill offered for fixing to powered harrows. Slowly and cautiously it creeps in here. I like the idea, but, we must face it, like the peg drum that could have bridged the gap between the binder and the combine, perhaps this is an in-between that will never get off the ground in the quest for precision.

I am impressed by machines such as the Amazone marketed by Taskers as a one-pass drill-cultivator-roller. It has a reciprocating harrow followed by a culti-packer, in turn followed by a 21-coulter drill. There is also the Lely Roterra, which lends itself to the combination better than most. Like Bettinsons with their drill, Sisis Equipment offer a rotary cultivator coupled to a direct seeding unit. Slots are cut at 100 mm widths and the seed delivery is controlled by rubber rollers which remain closed. Seed delivery occurs with the rate controlled by the speed and compression of these rollers. Both cultivator and seeder can be used separately if desired.

Even the centrifugal drill mentioned earlier becomes the 519 minimum tillage drill from ICI via their Plant Protection Division. Like the Carier, it uses what we would now call conventional spring tines. We really can ring the changes without moving far from the accustomed principles. The Ransomes-Nordsten Universal sowing unit can be called a combined drill or not according to taste. It drills the corn and sows the fertiliser from a box at the rear in broadcasting fashion. Another Nordsten, their grass seed drill, has coulters to scatter the seed over a width of 35 mm.

MASSEY FERGUSON '30' DRILL

The Massey Ferguson 30 drill can be 13, 15, or 20 row. It can be used for grain and fertiliser or grain only. As a grain drill it can have up to 31 coulters at the narrow spacing. As a combined seed and fertiliser drill it can be on the traditional wider spacing. Both materials are exuded from box to spouts by nylon-fluted rollers, thus both sections are using the same type of mechanism. The hopper on the combine drill has a variable partition to alter the grain-fertiliser ratios, if the individual farmer has reason to do that.

The MF 30 drill can have single disc, Suffolk, hoe, or tine coulters. You pay out your money—and it is not a cheap drill—and take your choice. The MF 30 has established for itself an excellent record of accuracy. The ram that puts pressure on the coulters is operated from the tractor hydraulics, though depth can be pre-set. Obviously, the MF 30 is not far removed from being a minimal tillage drill.

There are many good drills on the UK market, several from the Continent and Scandinavia. As in every other section, they are too numerous to mention and the variations of a theme are multitudinous. One that should be mentioned, however, is the Carier spring tine drill in conjunction with the Chafer saddle tanks for liquid fertiliser, as a combined operation. The applicator placed the fertiliser between every second row of grain and puts it 60 mm below sowing level. Tanks mounted either side of the tractor engine in true saddle style are of stainless steel, and a peristalic type pump is used to meter the fertiliser according to forward speed. The centrifugal pump used for spraying and filling is driven from the tractor's hydraulic system.

LIQUID FERTILISERS

Should we use solids or liquids? There are good arguments on both sides. I would base my own decision on how near or far the liquid fertiliser depot is from the farm. If observation serves me right, this is exactly what farmers are doing.

Drill variations extend into the province of fertiliser placement and granule applicators, such as the Horstine, to take a common example. Such applicators for granule insecticides and herbicides can be fitted to various framed implements. They can be fitted to precision drills, and potato planters, besides hoes used for inter-row work, or they can be used in their own right.

The broadcaster box for small seeds still stays in the market, the Hunt and the Alvan Blanch being examples. The Webb floating ridge drill is for precision drilling on ridges. The Standen tandem drill is for high-density drilling of beetroot, carrots, onions, or small seeds. Russell's of Kirbymoorside, who were one of the few to fill that gap between the ordinary drill and the precision drill for sugar-beet, still market planters for vegetable seeds, beans, maize and so on.

Farmhand go for a patented fertiliser placement system with their combine drills, covering 3 m and 4 m widths.

Apart from the Howard, the Lely and many versions of the Nodet drill for cultivator-drills, Krone have added the Amazone to their rotary cultivator to make what they call the Kronedrill. Carier have provided a very good alternative to most systems in vogue by offering two cultivator-drills, but one is for fertiliser. Both are based on the spring-time harrow and this is a cultivate-and-drill system that is not to be discounted. It is common to give two passes with the spring tine to get a seedbed. One pass can put on the fertiliser and the other the seed, if both versions are purchased.

The spinning disc fertiliser spreader is well established in our

34 Combined fertiliser-drill-spray operations with Deutz Intracs. Foreground: pneumatic fertiliser distributor. Middle distance: drilling. Background: spraying.

35 The Faun two-row potato planter; which is marketed in the UK by Ransomes.

farming operations. There is a hopper, a pto-driven spinning disc and a lever to regulate the rate of application. In some cases there may be adjustment of where the material falls on to the disc and which side the majority is thrown out. That way it is possible to go round the outside of the field and not throw any fertiliser into the hedge.

There is an inherent danger in all farm machines of adjustments that are not absolutely necessary. However convenient and admirable certain adjustments may be, for every operator who will get it right and—this is important—there are several who not only get it wrong but forget to make the subsequent re-adjustment. It is rather like increasing the rate of sowing in the drill for headland work. You then drill three-quarters of the field and return to base to tell the farmer that the grain supply has run out! Any person who commences farming and thinks that these things do not happen is naive to the point of being a fool.

Nor is it disrespectful to farmworkers to reiterate such a fact of farming life. Many of us, farmers and farmers' sons, have made plenty of mistakes. Let me state quite categorically that the spinner, albeit one of our most simple machines, is one of the most difficult to use. As I have said earlier, it is these facts of life that are making the tramline system of seeding popular. A school leaver could drive in the tramlines, but it needs a man of above-average intelligence—assuming intelligence to really be common sense—to fertilise a field with a spinner and have no subsequent strips. The light-coloured strips are underlap and the dark ones over-lap.

Disc speed should be constant and the tractor tachometer watched carefully. I have spent hours watching spinners from behind, besides walking miles inspecting the ground pattern. Do not be fooled. Even sizeable variations in distribution pattern are often not visible in practice. Looking on the ground is only slightly better.

The spinning disc was used years ago for lime, slag and other materials. In those days it was a trailed machine, of course. However, spinners soon went three-point-linkage mounted. That made them, in my view, much better. Addicts to indoor calibration should not get things out of perspective. Fertiliser of the types used today will bounce around; out in the field, it may not go so far as you think. To be fair to manufacturers, I must say that spinner performance has now improved greatly and the instruction books are better.

Much credit goes to Fisons for their demonstrations and for their tray system of checking. The series of trays across the full width of spread is one of the few methods of knowing exactly what is going on.

Evenness of distribution can be partly checked by spreading on a concrete floor. Take a fore-and-aft line down the centre, then sweep up the two halves and weigh them. But get the centre line right! After that, one has the makers' instruction book settings and the visual inspection after spreading some in the field. You wait for it to take effect on the crop and hope for the best. Often it is the worst! These days, since the tremendous improvements and the testing work done by such firms as Fisons, I would incline towards putting the instruction book settings as paramount.

Besides the spinning disc spreaders there are the pneumatic models such as Alpha-Accord and the Vicon. The unique one—at least it was for many years—is the Vicon swinging outlet machine. In hilly areas, both here and overseas, this pendulum action machine is often the most popular. This is not a coincidence. The NIAE in 1962, using the tray method of test to simulate slopes of 10 degrees, found that the pendulum machine was virtually unaffected in its spread pattern by slope. Where the results showed distribution pattern affected badly by inclines on the spinner types, there was no detectable effect on the machine with the pendulum outlet.

Spinners are no longer confined to the small mounted type of machine. There are the large trailed Atkinsons—the Hydramatic, with spreader mechanism hydraulically driven, is an example. The Tasker Paterson Fertispread is one of the outstanding successes in the machinery sphere. It can even have its own loading scoop. That way bulk fertilisers can be used, but farmers have refrained from rushing into bulk methods here. One supposes that the reason is because it is easy to calibrate fertiliser by telling a man to use, say, three bags per acre (as near as he can guess his acres). To send a man to put the equivalent amount on the land, but show him a heap of several tonnes to scoop up from is a very different matter.

FERTILISER INJECTORS

Taking the Leaper Ramsay as the specimen, there are now fertiliser injectors available. Such machines can inject liquid fertiliser, urea solution or aquammonia behind a series of thin knives. Because of climatic differences, these machines are more suitable in some countries than others. I have had mixed results, one year good, one not so good. The system is good on grassland, but in all cases I suggest that soil conditions at the time of application are most important. If the soil is not friable enough to fall in behind the slots, rolling can help. The ideal would be to inject the winter wheat nitrogen in the autumn, so that it is available for the spring. A sunny Christmas may alert it too soon. As always, we cannot rely on our weather.

The principle of pneumatic seeding is used in the sugarbeet sphere, alongside the belt feed drills like the Stanhay and the cell wheel as used by the Webb, but a word of warning: no matter how accurate seed is dropped onto the ground, if it rolls fore or aft, the precision is partially lost.

As has been mentioned with the new principles of grain seeding, precision comes very much from the short drop from the ejecting mechanism to the ground. Life has been made easier for the manufacturers now that we are near to being standardised on single-germ seed duly pelleted. The old, natural seed has disappeared, and polyploid is losing popularity. Monogerm seed is going to give the grower a 'doubles' figure of nil to, say, 5 per cent if the seeding mechanism of his drill unit performs properly.

Whatever mechanism is used—whether air stream, disc, cell, or belt—the method is to gather, convey and eject a single seed above the furrow opener or similar ironmongery in contact with the

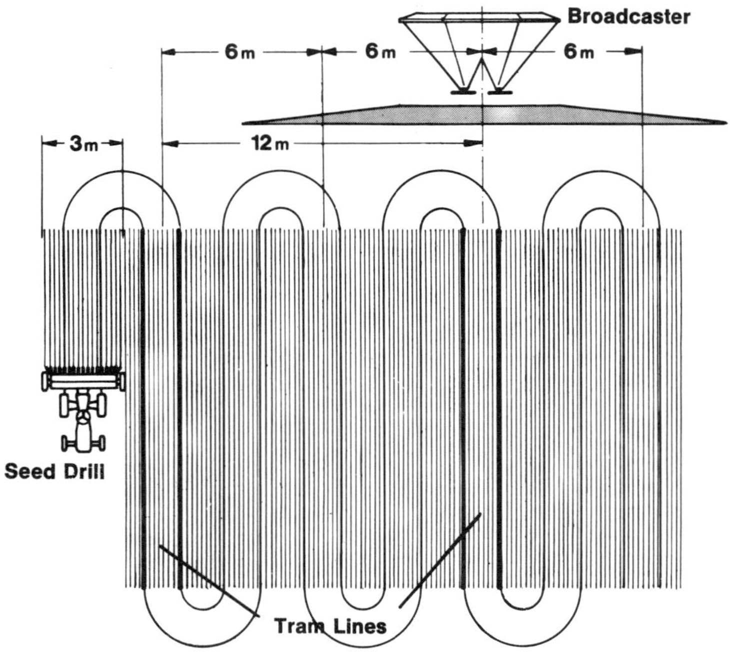

Creation of tram lines.

Ack: Taskers Trailers Ltd

36 This aerial picture of a Chafer sprayer at work illustrates the effect and advantages of the tramline technique.

37 Chafer's tramlining device uses compressed air from a cylinder to two pneumatic rams. The seed is cut off from two coulters every third run of the drill.

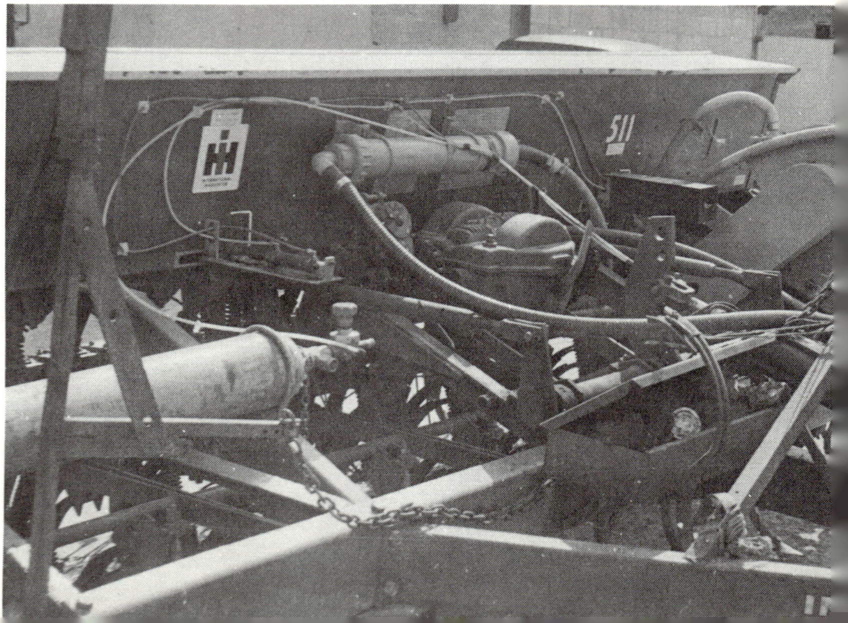

carefully prepared seedbed. No beet grower needs telling about the importance of perfect seedbed conditions for sugarbeet. Below par soil conditions are often necessary in bad seasons, but whereas cereals may make up over the season, rarely will sugarbeet come good at the end if the conditions were not good at the outset.

Where one may be able to roar a tractor up and down the field at 11 km/hr and more sowing cereals, the optimum for beet is likely to be nearer 5 km/hr. It does not matter that this speed has to be so sedate (provided the weather is kind) or if the land is 'over-drilled'. Beet drilling is a painstaking job, ideal for the person who is cool, calculating and even a bit lazy! The slow speed is partly necessary for straight, accurate driving if row crop work is to follow.

Insufficient research on finding an optimum for row width in sugarbeet leaves farmers very much to their own devices, and manufacturers at a disadvantage. Provided the farmer gets 75,000 to 87,500 plants to the hectare, it does not seem to matter very much how they are spaced, so long as they are there and not too irregular. In other words, beets will be closer together in wide rows and further apart in narrow drillings, to finally arrive at the optimum. At the moment farmers grow beets at anything between 460 mm and 610 mm.

Light-land farmers tend to grow their beet closer than heavy-land farmers because they want them to cover the ground against drought earlier in the season. Also, they are not expecting the beets to grow so large as the heavy-land beet, and the problems associated with the tractor wheels in between the rows, either for row crop work or harvesting, are expected to be less. The single seed drills now account for 100 per cent of the sugarbeet market. No farmer buying a new seeder for sugarbeet and similar crops, plus, of course, maize, is likely to buy anything else.

These drills are too well established and so little challenged that there is not a lot of enumeration to be done here. Even the battery-driven electric monitors, using a photo-electric element that not only keeps flashing lights on a console going for each unit, but also a minimum seed warning light, are standard equipment. It would not be unreasonable to assume that such devices, or the equivalent, could eventually apply to the grain drill or the ordinary drill. The belt feed drill is the one that seems to have captured the market in the UK so far as sugarbeet are concerned.

Sooner, rather than later, I am certain that makers of both drills and harvesters, to say nothing of inter-row machinery, would appreciate being able to design for standardised row widths, or nearly so. We cannot make a law about it, but it seems that all sorts of row

38 The Ugerlose front-mounted hoe from Denmark, which has met with much success in Britain.

39 The RDS Acremeter gives a direct reading of acres worked on the dial (top left). The circumference of the tractor or combine non-driving wheel is set on the three upper knobs on the right, the width of the implement on the lower set. The instrument can be fixed inside the tractor cab, within easy view of the driver.

widths come about because of local lore, the common width of wheel settings on the individual farm's tractors, or because of what suited the previous generation.

One certainty is that, whatever row width is chosen, or what is adopted as the optimum, it will have to be calculated arbitrarily, or as a sort of co-efficient between what is best for tractor, drill, inter-row implements, sprayer and harvester. That way, we could arrive at a figure round about the 510 mm mark. When I started farming I could only afford one drill. It was a combine drill with the common coulter width of 175 mm. Hence my beet were drilled at 525 mm centres because I blocked up two coulters out of every three! Before any beginner try to economise in that rather crude way, I warn that it may be that the two markers have to set differently. For good measure, I will add that the same drill top-dressed the wheats (and harrowed them) with the pressure released from the coulter springs. So I hope that nobody invents an all-crop, all purpose drill and thinks it is the first!

POTATO PLANTERS

Potato planters can be automatic or semi-automatic. The latter not only keep us nearer the traditional hand methods, but the human factor, unlike many other farm jobs, probably makes for more accuracy. Where chitted seed is used, it has to be remembered that the metal cups on the automatic cup-feed types may do some damage. Despite workers only being able to handle anything from 100 to 150 potatoes per minute, so that travelling speed is even less than for sugarbeet drilling, most farmers go for the semi-automatic machine. The operator's job of placing potatoes into cups or onto the endless belt type of planter is naturally boring and repetitive. With the tractor driver restricted to 3 to 4 km/hr, his job, like drilling the beet, is somewhat frustrating. Where semi-automatic planters are used, provided all workers are capable, some switching around is possible.

Potato planters are often combination implements. They can have fertiliser attachment, spreading or the more sophisticated placement method, besides ridging plough bodies. Taking as an illustration the Plantmasta from Richard Pearson, this machine can be bought in 2–4–6 row form, and standard equipment includes an electronic warning device; fertiliser and micro-band attachments are also available. Machines such as the Plantmasta and the Gruse from Root Harvesters, despite being automatic, claim operating speeds of up to 8 km/hr. Chafers offer a precision liquid fertiliser conversion for the Howard Rotaplanter.

The Hassia automatic planter has an automatic compensating device so that if a feed cup misses a potato, the roller control trips the mechanism on a 19-tuber reserve tray. The Smallford has a metering device delivering the seed in groups of six at pre-determined intervals to a conveying mechanism consisting of patented rubber caps. Two operators make good the doubles and misses, and this machine can have discs or ridging bodies, besides twin-band fertiliser placement unit, insecticide applicator, or bulb planting attachment.

Horstine Farmery have a fertiliser applicator for high speed planters that is front-mounted on the tractor and ground-driven by a spider wheel. Potato planters can be mounted or trailed. An example of the trailed variety is the Amac from Benedict which can carry 1,500 kg of seed; it plants four rows and has a hopper with a large surface area for bulk loading if necessary.

Ransomes have the Faun from Norway, which, like the harvester of the same marque, is designed for lightness. This planter has double cup belts, a micro-switch on top of each cup-belt assembly, and a feeler finger provides electrical planting control. A warning light for each row is in front of the tractor driver.

But the big development in potato planting is not really a planter at all.

I have long held the belief that the inherent British cussedness for pulling off miracles against all odds has done us no good in the sugarbeet or potato field. The trouble is that many of the kinder soils that are suitable for these crops are riddled with stones. Such stones find their way into every chain and sprocket in the harvester and are a nuisance in the spring.

THE STONE PICKER

So we have a new line in machinery—the stone picker. Taking the Geith, from Doggett (Engineering) Ltd., County Meath, Eire, as a specimen, a feature is that the machine is not power-driven in any way. A pneumatic road wheel is removed, using an ingenious jacking system, then the machine is lowered to bring a large steel wheel in contact with the ground.

Rakes collect the stones and move them into pockets in the massive, rotating ground-driven wheel. Stones are elevated into the skip. Collecting rakes are hydraulically operated for efficiency on undulating terrain and the steel hopper is tipped by hydraulic ram. This particular machine makes claim to picking stones up to 450 mm diameter and is widely used in civil engineering.

Another method of dealing with stones in the potato field (introduced by Pearson's of Boston) is to ridge, fertilise, plant and also pick stones in one pass. Yet a further idea is to fit the stone picker to the harvester, but, one suggests, getting rid of the stones at spring seeding is better. Soil conditions would be better than they would in an unfavourable autumn and, as with beet, there are enough harvesting problems without any more gadgetry added to machines. At the planting end, the method is to pick up soil and stones, the latter remaining on the web. Then they get discharged sideways to a ridge bottom. The discharge point is re-positioned hydraulically for each run across the field. The baulks have to be good, so that the row of underground stones remain undisturbed by the harvester.

Either of these methods work, but with the real, purpose-built stone pickers there is the added advantage of having some stones to use for gateways and tracks around the farm. Failing that, if they are not wanted, there must be a market for them.

SUMMING UP

To sum up. So far as corn drills are concerned and despite the price, one has to say that the Massey Ferguson '30' range of drills are unsurpassed. After that, I like the International drills and the Nordsten. In any value-for-money class and for going straight onto the furrow, apart from a combination unit volving a cultivator such as the Lely Roterra, I would suggest the spring-tine Carier. It does a fair job at a fair price.

In the sugarbeet field, it has to be the Stanhay with belt feed. In the matter of spinner fertiliser spreaders I like the Amazone, except for hilly areas where the Vicon with pendulum action scores. If I wanted the biggest and best with no financial strings, either for big-farm work or contracting, there is one machine on our market that I have a tremendous respect for—the Big A brought from the US by Bunns of Great Yarmouth. For what this machine will do, its price is nowhere near as exorbitant as may appear at first sight. For a big-time spreader but tractor-drawn, as against the Big A which is self-propelled, there is the massive Amazone. For the ordinary sort of trailed spreader at medium to large farm size level, there is the Atkinsons or the Tasker-Paterson.

Chapter 4

FARM SPRAYERS

IN THE early days of farm spraying, someone coined a phrase 'In May let us spray'. It is true that in the early period of spraying by the farm staff, as against the professional contractor, May was usually the ideal month for spraying MCPA around the countryside. While this May-spray dictum is still apt, the farm sprayer now has a universal application that takes in about eight months out of twelve.

Taking the average sort of arable farm growing the average crops, and leaving but a month for harvest and the three winter months, means that the sprayer can be in action for two-thirds of the year, as against a combine harvester's one-twelfth. Farmers incline to be a little mean about the sprayer only because everything connected with spraying was so cheap in the first instance.

Many of us have had to spend a proportion of our youth walking the rows of barley chopping out thistles, or spading docks from the wheat field. Just as many of us must have prayed that some miracle would come about that prevented the growth of anything other than what the husbandman wanted in that particular field. This is exactly what has happened.

INSTANT WEED ELIMINATION

The withering weed plants, one day after spraying, would have appeared more of a miracle to our forefathers than anything else. What made the new situation so pleasing to the farmer was not only the miracle of instant weed elimination, but also the cheapness of it all. In 1950 one could buy a sprayer for less than £100 and get rid of the common weeds in cereals for a few pence per acre. The whole concept of chemical weed control was to become the most absorbing of farm topics.

Of course, we had not even started on sprayers and liquids. We had a short spell of a year or two on powdered chemicals spread on the crops with a fertiliser distributor. Powder was not a failure. Applying it was not the nicest farm job if the wind was wrong, but until tractor cabs were introduced, the same drawback could apply to liquids.

We now have everything from small knapsack sprayers to self-propelled, purpose-built sprayers. Funnily enough, despite the British farmer rushing headlong into self-propelled combines, almost to the exclusion of everything else, we have been a little slow in moving to the purpose-built sprayer. Even after such machines were to be seen at Continental shows, we appeared to be somewhat reticent.

BIGGEST COST-SAVER

Nothing should stint our praise for the machinery designers and chemical manufacturers who did so much in this important evolutionary phase in post-war farming. Arguably, the farm sprayer could be rated the biggest cost-saver of recent years. We tend to be a little scornful of those who have benefited so much from the spraying revolution. The makers of such equipment had a good time and the chemical manufacturers enjoyed a real bonanza. At the end of the day, it was the farmer who benefited most.

It is true that there are times when the farmer justifiably feels that he has been ‘taken for a ride’. A new chemical to deal with an old weed problem will always be expensive until the initial research has been paid for and until a volume market exists. This is nothing peculiar to agriculture. It would be the same in any other industry.

Such price rises have mainly been something that nobody in the UK could do anything about. Admittedly, with a new chemical and in times of high unemployment, there have been occasions when it looked cheaper to get a gang of men in to pull, say, wild oats by hand. This is always to assume that men are keen to do such jobs and that the farmer is willing to tolerate the degree of supervision that would be necessary. Of course, if the cost of machine and spray is too high, there is nothing to prevent a farmer resorting to the old methods if the labour is available.

Weeds, it is said, choke 10 per cent of food crops on average. When a sprayer goes into the field the crop damage done by the machine can be 5 per cent, so any crop benefit has to be well above that. Even in the advanced agricultural areas of Europe, North America, Asia and Africa, losses could well average 25 per cent. Total elimination of weeds, pests and diseases by the machine-chemical combination, we were told, would result in sufficient food to feed all hungry people in the world. It will never happen.

For as far ahead as any of us can see, the Utopia of an agriculture needing no antidotes is not to be. We are warned that agrochemical prices will go even higher, and that some boom and nozzle combinations are so inaccurate that the dosage rate can vary by a factor of up to a multiplication of 40. Still, farmers will spray in May or any other month that suits them! Nobody has the right to castigate them for doing what they know to be right.

We spray herbicides for the well-known reasons of reducing weed competition and increasing yield, besides easing the harvesting operation. We spray to burn off potato haulm. We use paraquat and diquat, especially in connection with direct drilling. We spray insecticides and fungicides, besides applying liquid fertilisers.

A wettable powder must be agitated because it will form a suspension in water. A fluid will form an emulsion with water and may or may not need agitating. A fluid for mixing with or soluble in water, or a soluble powder, will probably need no further agitation. Agitation can be important, or it can be a nuisance if frothing takes place.

WILD OATS—A SERIOUS PROBLEM

So far as arable farming in East Anglia is concerned, with wild oats posing one of the biggest problems, the sprayer soon became established as one of the most important pieces of machinery on the farm.

Most of our spraying is low volume through nozzles with fine holes. Coarse suspensions need larger nozzles and higher volumes. Nowadays, many farmers use liquid nitrogenous fertiliser for top dressing and a few use NP and K compounds in liquid form. My experience is that the ordinary, low-priced farm sprayer does not take too kindly to nitrogenous fertilisers. Such corrosive materials attack ferrous metals, and possibly the additional expense of stainless steel is justified when liquid fertilisers are used.

As the years go by, one supposes that there may be more high-volume spraying with still more sophisticated sprays, so that there is a need for more proficiency in the farm sprayer.

THE MOUNTED SPRAYER

Yet, as with all machinery, if one can get away with the simple, low-cost, low-volume sprayer, there is no reason why it should not fulfil its function well. I like the manoeuvrability of the mounted sprayer. Obviously, now that sprayers are getting larger, there is often a need for a trailed type, and such firms as Chafers, the liquid fertiliser people, usually use that type for their hiring service.

For small and medium-sized farms, I do not think that the mounted type can be bettered. Its ability for getting into corners of odd-shaped fields, or for setting in at the headland if the tractor wheels have to be arranged to suit, for instance, rows of sugarbeet is something that one partially loses on the trailed type.

The Chafer sprayers are good and the hiring service is something not to be discounted. The same remarks can apply to the Allmans and Boots sprayers.

One good thing about hired sprayers is out-of-season maintenance. The sprayer hiring companies usually take the sprayer back at the end of each season and thoroughly overhaul it. This also means that an annual hiring agreement can be terminated conveniently if a farmer wishes to change to another company, or purchase his own machine.

There are, by now, newer and more selective sprays that kill one or two more weeds, besides the traditional docks, thistles and charlocks. Even if the farmer has not got any of these other weeds, he could well have been talked into the new, but more expensive chemical. After all, he had got to be in the fashion! The chances are that he would probably, especially if on heavy land, have been better to have spent the money on wild-oat killers, even if they did look expensive.

Cost and cost alone kept many farmers off wild-oat killers for too long. Even now, many still use the simple and cheapest of weed killers, and in some cases in the cereal field, none at all.

LACK OF MARKER SYSTEM

The main shortcoming with the technique of spraying has always been the lack of any marker system. As with so many field operations, if in doubt, go round, not up and down. That way the mistakes are not so well telegraphed to the outside world! The new technique of tramlines in cereal fields, is, one supposes, a way of getting things right. I am not over-enthusiastic about tramlines, despite the growing popularity of the system, but more about that elsewhere.

From dye markers to a bag on a stick, I have seen few systems that work well. The dye or blob system as it gets called uses a small pump to give series of blobs across the field that look rather like a housewife's detergents, but it works well. The foam is contained in a canister and pressurised out, while dye is normally pumped either continuously or intermittently. Just one word of warning about the blobs of foam. Do not expect them to be waiting for you next morning, nor even after a long lunch. They may stick around for about three-quarters of an hour, but not much more.

White sticks, or old fertiliser or feedingstuffs bags on poles, often set out over the whole field before work and collected after work, are still much in vogue.

Most of us have tried the trailing piece of chain at the end of the sprayer booms, or the weight on the end of either rope or chain. The wider the boom, the more difficult it is to see the mark. That, of course, applies to all marker systems. A cycle or motor cycle wheel on the boom end is another similar system. The trailing piece of string or rope works if everything is true and straight. Incidentally, if the work of the drill is up and down, do not forget that the accuracy of the drillman will be reflected in the spraying if one relies on his work.

A long piece of string on the boom ends will trail straight—always provided the tractor driver is driving straight. Do not be fooled, not all of them do! On turning, of course, the operator has to keep his inside boom end in more or less the same place. Thus when he commences his return across the field, he can start off by watching the inside boom end trail directly over his piece of string. After that he is on his own, but if the drilling is accurate and he keeps his tractor wheels on course, he should be all right. It has one big advantage—it costs virtually nothing.

I think that possibly the ideal marker is that fixed to the front end of the tractor. The trouble nowadays is that sprayer booms are often 12 metres wide. This takes a lot of metal, a lot of weight, and, of course hydraulic power to raise or concertina the booms in and out. It has always been my hope that some of the tractor manufacturers might supply marker systems of one kind of another with their tractors. At present we have all the people in the drill, sprayer and fertiliser business trying all sorts of systems on all types of machines. One satisfactory marker unit affixed to the prime mover would suffice for all subsequent operations requiring any marking.

Apart from drainage machines and the like, we cannot think as big as laser beam systems. Back in the 1960s I was shown a scientific system at one of Fison's research centres. They had even made their own sprayer to go with it. I did not see it again.

For tomorrow's farming, Fisons were on the right track. Similar systems using radio control are under review. Between the sticks and radio control comes the idea of a series of mirrors. For some time, the NIAE have experimented with mirrors in the form of a horizontal periscope, with one mirror on each boom end and one on the tractor bonnet. A mark at the boom end should be reflected onto the mirror that is in the driver's line of sight.

40 The Everall 800L tractor-mounted agricultural sprayer with an 800 litre capacity tank.

41 Ransomes 12 metre sprayer boom for metric sprayers—the Cropguard 700 Super and 500 Super.

UNDERLAPPING TROUBLE

I have attached some amount of importance to markers not because of all the new-found interest in the economies of a bit of overlapping, but because of the trouble that results from underlapping. The telltale strips of charlocks or wild oats are galling to both farmer and operator. One of the problems of wild oat control by Avadex has always been the unfortunate need for incorporation, but the advantage, nevertheless, is that working on bare land, as against a growing crop, the chain, weight, or wheel marker system will work. The tractor wheelings are usually visible to help give a guide.

On a growing crop of wheat or barley, it is a very difficult proposition. With sugarbeet, potatoes, maize and so on, of course there are few problems, for the obvious reasons of having the wide-spaced rows to drive the tractor wheels in. Sprayers are another branch of farm machinery where there has been some uniformity introduced, so that marked differences in quality of manufacture, or field performance, are not very prevalent. The old gear pump, susceptible to wear when using suspensions, is not out of fashion. Piston pumps can generate tremendous pressure but may be too expensive for the agricultural market, apart from specialised machines for special crops. Centrifugal pumps are much cheaper and need little maintenance, but they are more suitable for high-volume machines and are much less of an econoimc proposition on the ordinary type of low-volume sprayer.

ROLLER VANE PUMP

The roller vane pump has been the one to catch on but it is giving way now to the diaphragm pump. The roller vane pump is a single rotor on a spindle mounting which is offset from the casing. The rotor has a number of rollers that are made from various substances, such as rubber, nylon and nylon-coated steel. In action the rollers are thrown into contact with the casing, and as the rotor is offset, the rollers move in and out radially with each revolution. At their most extended position, the fluid is drawn in and carried between the rollers and casing and forced out of the outlet as the rollers are being pushed inwards.

The liquid is drawn into a chamber by the diaphragm which ejects it with its to and fro flexes. Delivery and suction valves control the flow of liquid. Fairly high pressures are obtainable from the high speed small displacement diaphragm pump and for farm use (though not everybody would agree) it has the sudden death advantage.

Where a roller vane pump gradually becomes less efficient through wear, if valve or diaphragm failure occurs in the diaphragm pump, then the changes are that the operator will have to report it as unserviceable.

This is not as bad a principle of machine operation as may seem at first sight. I am all for 'bodging' to keep going, if no harm is done. In fact, men capable of in-the-field 'bodging' are often invaluable. But spray pumps are not items for such treatment. They are one of those things that need to work well, or not at all. The plastic tank is increasing fast in popularity, so that the problems of corrosion with ferrous metals mentioned earlier are becoming less applicable. I like the plastic tanks, especially the rounded edges which do away with sharp corners on the exterior and recesses on the interior.

In choosing a sprayer. I think it fair to say the the farmer can safely leave the technicalities of pump type to the manufacturer and concentrate on the type, size and price that suits the purpose for which he is purchasing.

SELECTING NOZZLES

Nozzles come and nozzles go in the advancement of spray techniques. I used to use a sprayer where we had one particular type of nozzle, then came one supposedly better, and another, and possibly another. Then, hey presto, we were back to the one with which we started! To be fair, I am certain that the ever-changing cost of raw materials has something to do with it.

Nozzles can be the expensive stainless steel, ceramic, brass, nylon or plastic. Again, as with overlap, we are hearing more about the efficiency or otherwise of nozzles these days. Most nozzles are turned out to be accurate to within plus or minus 5 per cent. Brass will wear quicker than most of the others and no doubt that is why many farmers rejected it.

If several hundred hectares are sprayed in a season, it will hardly affect the overdraft figure at the bank to have a new set of nozzles each season. After all, a man hoeing sugarbeet usually treated himself to a new hoe to affix to his old handle each season!

IMPORTANCE OF BOOM HEIGHT

By now most spray users should know the importance of boom height above the crop to avoid excessive overlap from nozzle to nozzle and equally to avoid missing strips. There is also the matter of spray drift. Apart from not spraying on windy days, obviously boom

42 The Dorman Super TM4 mounted sprayer incorporating the Dorman 'Regulo-Flo' system of application.

43 The Dorman 680 litre saddle-tank sprayer with 18 m spray boom.

height has to be set at the optimum. Nozzles may be cone or swirl type or, more likely, the fan type. Vibrating nozzles and atomisation by air blast are features now with us. Too much atomisation could induce unwanted drift, perhaps onto a neighbouring farm.

As with the tanks and pumps, farmers do not have to concern themselves overmuch with nozzle types. I have always been a little critical of the ceramic-tipped type that chipped so easily if they strike the ground or some obstruction. Is is very necessary to have plenty of spare nozzles in the tool box, plus the inevitable piece of fine wire. I know that the piece of wire is frowned on, but I reckon it is not easy to do without it.

It is more important than some operators realise to clean the filters on the sprayer. Maybe some will assert that filters are not so accessible as they should be, but it may be more valid to say that it is not always easy to get a good deal on some models of sprayer after removal and cleaning. Spare rubber seals are, in some cases, I have found, as necessary as spare nozzles. If drift is to be avoided, it is necessary to correlate nozzle size and pressure to a pretty fine art. Pressure gauges also seem to be a fairly fallible piece of equipment. Like the nozzles, they need an annual test and quite often need to be thrown away with the nozzles.

MOUNTED WATER TANK

While the self fillers that most sprayers have on them are excellent if the correct sort of water supply is at hand, I have often found that saving a journey back to the farm by drawing from a pond or river can be more costly in time. Apart from blocked nozzles, it will often be necessary to do much more filter cleaning or even filter replacement, which can result in a net time loss. The importance of clean water cannot be over-emphasised. Most farmers have progressed from the self filler to a more suitable arrangement. Usually this is a tank with ball valve filling set high on a stand, with a tap opening of, say, 75 to 100 mm and which fills while the driver is in the field.

With that sort of layout, the operator can fill his sprayer in minutes, and now that sprayer manufacturers have got around to sensibly sized hatches on top, it is no great danger in putting the chemical in as the tank fills. Despite anti-drip mechanism and 'suck-back', many sprayers seem to manage a drip at the place of filling. When we refer to spraying dangers, this is one that could deserve more attention. Assuming a natural standing, as against concrete with suitable run-off, a risky concentration of, possibly, several chemicals must occur if a permanent filling place is used.

It is common for farmers to mount tanks onto trailers and cart water to the field. There are circumstances, especially if a second man is available with nothing better to do, when this is advantageous, but quite often I cannot see any difference between carting the water to the field and the spray operator coming home for it. Sprayer manufacturers in their instruction books now give a guide on how to calibrate the sprayer in question. Often a chart is printed on the sprayer itself for quick and easy reference.

I know that calibration of the farm sprayer will sometimes have to be done at farm level, but a lot of frightening talk about calibration is not good for the user. Farmers should not assume that the makers did not calibrate the prototype! This applies to sprayers, seeders and fertiliser spinners. Admitted, the early models often left a lot to be desired in accuracy, but the farmer, on the whole, would be better advised to tell the operator to stick rigidly to the makers' chart and instructions rather than be clever and do his own calibration.

MANY SIZES OF SPRAYERS

Sprayers are no longer the prerogative of the professional artisan. From large estate to small-time husbandman, all types of sprayers and almost every conceivable size are available from the new self-propelled monsters to a knapsack sprayer for the gardener. ICI, through their Plant Protection Division, even offer the Aquatic Weed Control Boat. This is a self-propelled boat with engine-powered spray mechanism for water weed control, plus insecticides.

The Hardi Pick-Up Unit Sprayer has a petrol engine and is designed for placing on a trolley or for putting into Land-Rover, van or pick-up type vehicle. The Turbair Sprite is operated by a 12-volt battery (or mains or petrol engine) for ultra low-volume spraying. The little knapsack sprayer, where the user pressurises by hand action pump, is a real boon for work around the farmstead or even for strips that get missed on a field scale.

I have found the hand lance, usually offered as an accessory with farm sprayers, to be a very worthwhile purchase for jobs that can also be done with the knapsack sprayer; it is useful for buildings too. Even the little pto-fitted roller vane pump from Lawrence Edwards & Co., that is mainly used for washing down machines, can do small-scale fruit tree spraying.

The band sprayer was favoured by sugarbeet growers, then it lost ground as chemicals got cheaper and farmers went for overall spraying and less inter-row work, but it stayed the course.

Never a great fan of band spraying, I felt that it was a case of swings and roundabouts. But this is far from the end of sprayer models. Now we have separate band sprayers for front mounting, like the Dorman. Or we can have the Broadwater that fits on the front end of the arms of the Ugerlose front hoe for pre-emergence, post-emergence and into-row application. Then, for example, the Allman 'Intaro', using Plant Protection's 'Vibrajet' nozzle system.

IMPORTED TRAILER MODELS

Large trailer sprayers have come to our shores to supplement and compete with the home-produced models. The Evrard from France and the Caruelle and Berthoud ranges are examples. The Warnock is a hand-operated inter-row crop sprayer, three-point-mounted, with individual plastic containers for each row and a seat for the operator above the tractor driver.

Many farmers now fix a boom and nozzle bar to a cultivator for spraying, for instance, triallate, which has to be incorporated quickly owing to being somewhat volatile. Others put the boom onto the front of the tractor. We used to have a boom for the front-end loader, with the tank mounted on the linkage at the rear.

Besides the front-mounted types for on-the-row, between-the-row and all types of application, Evers & Wall offer their rear-mounted 'Tween-Row' for the same job.

Such firms as Fisons, Chafer and Dorman use saddle tanks mounted on a framework so that there is a tank either side of the tractor engine. Both front and rear are thus free for the boom.

Little gadgets that do a lot are exemplified by Allman's chemical self-filler attachment. The operator puts the chemical into a pail or some other receptacle so that the attachment can draw the chemical into the sprayer from ground level. This any danger from the chemical is eliminated.

Another spray variation is the underleaf method of upwards-facing jets. Yet another is to band-spray and inter-row spray with two materials from two tanks. In other words, taking sugarbeet as one example, one could be spraying, say, gramoxone between rows and betanal on the row.

INACCURATE MARKING

While it would be difficult to find people to admit it, one of the reasons (and it could be paramount) for the arrival of the tramline system in cereals is because the average farmworker is not good at matching swaths at spraying time. No marker system is as good as the

44 The model 300 sprayer from E. Allman & Co Ltd fitted with a boom winch.

45 The Allaeys self-propelled sprayer.

skill of a good driver, unless it is the tramline system. Even when pacing out the bouts for the basic stick system of marking. It is surprising how many workers will walk at the angle of the boundary, rather than straight. With an angled boundary fence at one end of the field and a straight edge at the other, it is easy to see that the distances arrived at would be different.

The problem of inaccurate marking has been augmented by the increasing width of booms. It is difficult to travel a lot faster because of boom bouncing, which will occur unless the sprayer is fitted with a boom damper system of some kind. Although there has been a move to large trailed sprayers simply to get more capacity, I do not think that this will become a permanent feature. Of course, the larger the tank, the less running to and fro for water. But the answer is more likely to come from controlled-drop sprayers, atomising with a much reduced water usage. Less travelling, less filling time and less weight to carry on the land would provide the ideal.

The self-propelled sprayer is available in the UK in the form of the Belgian models from Allaeys and Delvano. The Delvano is built around a Ford 4000 or a Massey Ferguson 575, while the Allaeys is fitted around a Ford 5000. There is also the Sands built around a David Brown 990. These models have wide booms, large tanks, high clearance and wheels that should do minimum damage in the crops.

Always a difficulty with imported sprayers is the calibration if, after the initial set of nozzles has worn out, a change is made to British nozzles. Often recalibration is necessary, but it should be said that calibration is usually done with water in the first instance. It often happens that a certain amount of either recalibration or trial and error has to take place at farm level.

If calibration has to be done, it is a matter of calculating distance of travel at the given boom width to cover a hectare. Fill the tank to a pre-arranged level mark, spray the measured area at the recommended speed, then refill to the mark, measuring the quantity of water needed. Often nothing more than a slight variation of forward speed is necessary for correction. Most sprayers can be cleaned thoroughly with a 10 per cent solution of washing soda.

As with drilling, most operators spray the headlands of the field first. With large fields, it is inevitable that the tank empties during the run across the field sometimes. Many farmworkers, I notice, keep going and guess where the tank ran out, or they stop and put a stick up. This wastes time. Better to turn onto the sprayed part of the field, so leaving a cross truck to go back to, or conversely, if the filling point is across the field on the unsprayed side, I cannot see any great objection to running across that part of the field.

SPRAYING BY AIRCRAFT

A word about spraying by aircraft. We have only reached a figure of something like two million acres being sprayed from the air, or around $7\frac{1}{2}$ per cent of what, for instance, the Americans do. In Australia it is common to see heaps of phosphates awaiting the aircraft. As a matter of fact, 50 per cent of the operators working for the twenty-two companies using some ninety aircraft in this country are from Australia or New Zealand. Aerial spraying has increased with the arrival of antidotes for aphids and late-developing diseases in wheat.

It costs more to spray from the air and this has to be weighed against the wheel damage done by the ground sprayer going through the crop just before harvest. Not only that, but not many farmers have got round to owning their own aircraft or helicopter. One relies on a contractor almost 100 per cent. Often it is as well to tolerate the wheel damage that looks as though it must outweigh yield gained. The same sort of reasoning applied to air spraying, where the cost often looks to be near the worth of the extra grain harvested.

One cannot completely discount spraying from the air, despite the problems of drift and the alienation of non-farming people who live in our countryside. To be fair, such people did not buy a country cottage to be fanned from the air with chemicals which they do not understand. Dangerous or not, I concede that it can be frightening.

WHICH TO BUY?

In sprayers, I like the Allmans, plus the Allaeys-Allman. After that I tend towards the Ransomes, the Dormans and the Evers & Wall. If I wanted the elite in specialised, big-time sprayers for an all-in chemical and fertiliser programme, I would go for the latest in the Chafer sprayers.

Chapter 5

THE COMBINE HARVESTER

It took until 1928 to get the combine harvester working in British fields. The Americans and Canadians would rate us a little slow off the mark, but it has to be remembered that these were horsepower days and three-horse binder teams were about as much as we wanted to handle. I doubt whether horsemen would have relished the idea of 16-horse teams to haul this new combination machine around our comparatively small fields.

We were into the 1940s before we turned the thousand mark in combines and nearly to the 1950s before we had 10,000 of them. Yet we managed to surge upwards beyond the 50,000 mark by the 1960s. As with tractors and so many farm machines, figures are now pretty static. Advances are now in the form of size and technology, with numbers, if anything, going down.

The advent of the combine harvester could be rated the most pleasing aspect during the evolution in farm machines. For the sake of nomenclature, it should be stated that such countries as America, Canada and Australia call combine harvesters 'headers', while referring to combined seed and fertiliser drills as 'combines'. In UK we settle for the term combine in relation to the harvester and combine-drill for the seeder.

GREAT MECHANICAL ADVANCE

I always reckon the demise of the old threshing machines (stationary) in favour of the combined thresher (portable) to be one of the great mechanical advances of our time. My regret is that the previous wretched system survived the 1930s, to keep alive that dreaded chore known as 'chaff and colder' that was the bane of every school-leaver in the country areas! Replacing the puffing steam engine with what are now our vintage tractors did little to alleviate the drudgery allied to the binder and thresher method of harvesting cereals.

Cutting, stooking, carting and threshing the harvest was a laborious and ridiculous exercise. Foolishly and wrongly, farmworkers were quick to condemn the new methods as wasteful, conveniently overlooking the bushels of grain that were to be found at the bottom of the waggon. Or the heaps of grain and seeds mixtures that seemed to be the trademark of the thresher.

The farmyard hen covered a multitude of sins by scavenging what the traditional methods would otherwise have caused to be wasted. Even now, we have not fully appreciated the fantastic change. In the old days it was necessary to use blocks to set the threshing drum at spirit level accuracy. Hours were spent in muddy stackyards perfecting this necessary levelling operation, to ensure complete threshing of grain from straw.

Suddenly, spirit levels could be jettisoned, along with binder and waggon, and a machine that combined all operations romped round the field at all angles, levels and elevations with the crop in a green, immature state, churning out grain in one direction and all waste in another.

Admittedly, the grain was hardly the perfect sample for showing at Mark Lane, and the problem of how to deal with wake of loose straw was causing concern. But the great grain revolution was under way. The reaper, the binder, the threshing drum and straw pitcher were living on numbered days. Not only was their borrowed time to run out quickly, but every forward-looking farmer should have been doing all in his power to accelerate the acceptance of the combine harvester.

ONCE-OVER OPERATION

Over and over again, the traditionalists were heard to say that the first wet harvest would find the combine out. Alas, their predictions were as doomed as the sickle and flail before them. It was the complete opposite. The harvest of 1946 was to be a grim and muddy affair in the main cereal areas of Eastern England. The War Agricultural Committees that were still with us were about to turn again to the Western Hemisphere for help in the form of peg drums. These would have been a good in-between had we awakened a few years earlier. But a few farmers had tasted success. They had divulged their secrets to friend and neighbour. It was the completeness of a once-over operation that settled for all time in the UK any argument about the correctness of combine harvesters for our conditions.

In 1946, farmers with ground-driven binders, and those who were a little old fashioned with their tackle, struggled for two months and more in quagmire conditions. Those who had progressed to a power-driven binder fared better. They were not to be found prone on their backs wrapping sacks round the giant driving wheel mounted amidships on the old binders. Like everybody else, the lucky few who had a combine could only snatch at the odd few hours of sunshine that came but—and this was the crux—what they reaped was theirs for the keeping.

The binder men stooked, slithered and carted, while the lumbering threshing monstrocities awaited their inevitable call. The combine men soon found the snags with damp, often immature grain. In those days, our combine harvesters were baggers, so even this was not too big a problem. Sacks left open at the mouth and stood loosely around the old-fashioned buildings that most farms had at that time took little harm at 18 per cent moisture content. Even grain upwards of 20 per cent moisture survived with a bit of re-sacking and airing.

These were trifling exigencies compared with the binder man who had only done one part of the operation at every snatch. It was inevitably this completeness, barring the not very valuable straw, that appealed to the farmer of vision. By 1947 the combine harvester was home and dry. Every British farmer worth his salt should have assimilated the lessons of the 1946 harvest, that a grain in the barn is worth two in the field.

The peg drum came and went, but never forget that its history would have been so different in the UK had it been introduced years earlier. Binders were left to rot, rust and decay. Steam engines were left for enthusiasts only, while their accompanying drums and straw elevators were soon to be seen in the corners of stackyards, yielding to the elements and overtaken by the natural growth of grass, nettles and sheeps parsley.

INTRODUCTION OF TANKER COMBINE

It took several years for drying, cleaning and handling equipment to catch up with the combine era, but it did not bother farmers overmuch while they still used sacks. Soon, however, bulk was to take over. The bigger combine required two men, whereas the tanker got away with one. There was an in-between period in which farmers used a tanker combine with the second man with a trailer at the headland, where he held the sacks for filling from the combine spout. He would tie them up and arrange them on his trailer, besides unloading back at the farm, while the combine operator continued to his next tank full.

Apart from a bit of time wasting for the emptying, this was a better system than throwing down three bushel sacks of corn all over the field for some poor devils to hump at the end of a tiring day.

By about 1950 the grain-growing farmer could see the necessity for a smooth, automated harvesting operation with bins, or other bulk storage, plus some sort of drying facilities and aeration of grain stored in a mass. During the next decade, every contrivable means to that desirable end was to be tried out on almost every sizeable cereal farm in the country. Few systems were complete failures. It was just that some were better than others. No NFU meeting, nor its bar talk afterwards, was complete without some discussion about blowing hot air, cold air, drying by batch, drying continuously, round silos, square silos, indoor silos, outdoor silos, grain pits wet, grain pits dry, sucking grain, blowing grain, dressing grain, pre-cleaning it, conveying it, and, if any time was left, maybe someone would remember the little matter of selling it.

The corn merchants were left trailing. It was not until the farmers got heavily involved in grain production that they realised this was their trade that was being dealt with. Most of the country merchants were pathetically slow to get into the swing of things, otherwise we would have had a much better organised and better documented cereal trade than is the case. The local corn hall was very much in vogue. Merchants were too busy playing cat and mouse with every corn sample.

The great psychological act ruled their hearts, heads, hands and bank balances. First question when a sample was offered was to ask, always provided the truth would out, whether the purveying farmer had a drier, storage or money. If he had all three, the merchant was on a loser. If he had no drier and it was a wet harvest, the merchant would really call the tune. Lack of storage would bring the same ridicule towards the grain being viciously manhandled, criticised and cut.

MONEY PROBLEM

Shortage of money can work both ways. If the farmer has a large account outstanding with a particular merchant, this can be an incentive for a good offer for the grain, whether the farmer really needs money or not. Farmers are no less astute with their own brand of psychological trade warfare and it is not unusual to let the bills pile up before harvest, despite being in a position to pay. Kidding the merchant into buying grain to get his money was not, and is not, such a bad tactic.

Unfortunately, while the country merchants played ducks and drakes with their farmer clients, they were being small-minded to the extent of falling behind. Had they set about building stores and taking in the grain being hawked around the corn exchanges by farmers who had rushed into combine harvesters, small-time corn storage depots may never have mushroomed in the way that they did from 1950 to 1970. Despite the knowledge that was available in America, Canada, Australia and even amongst the co-operative stores on the Continent, it took a little while to persuade merchants and compounders that their hour of opportunity was at hand.

Just in time, merchants increased their drying capacity and buildings sprang up. When it happened, it stood us in good stead later Many argue that the money poured into farm improvement schemes to finance thousands of hotch-potch grain installations would have been better spent on communal grain stores. In that way documentation would have been assisted.

GRAIN STORAGE

Grain stored on the farm for farm consumption is so obviously right. Grain stored for sale is a very different proposition. There is no reason why such grain should not go from combine to store, as in the prairie lands, even if in our case it is for home consumption and not export. This could have given us a nucleus of data regarding stocks. As it is, our knowledge has to be part-data, part-guesswork and partly based on statistics which can get adjusted to suit the financial year on the individual farm.

These are the problems that came in the wake of the combine harvester, but statistically they are no more inaccurate than guessing the amount of cereals in the stackyards. Statistics became important because there was no longer the orderly, gradual marketing of cereals as was the case when the old threshing sets travelled around the farms between harvests. It was not easy to sell grain in the stack, but it was always available as a saleable commodity, once into sack or silo and always provided the moisture content was at a safe level of 16 per cent or below.

Combine harvesting was such a complete operation that few people, certainly not many administrators, realised what a transition had taken place. Marketing became chaotic and it was mainly the farmer who lost, though the taxpayer did not get a very good deal either. With cereal deficiency payments in vogue, not enough of the farmer's income was coming direct from the market. At least events proved to cereal farmers the value of the safety net contained in the 1947 and 1948 Agriculture Acts.

All this does nothing to detract from the genius of those who gave us the combine harvester. Our failure to deal with the aftermath of chaos in the market, or more to the point, our failure to foresee inevitable events brought troubles that we need never have had.

Not only is it remarkable that the combined reaper-thresher could cut, thresh and deliver at all angles, but also that the conception has changed so little. Bearing in mind that all combines are travelling threshers connected to a standardised cutting mechanism, it could be that certain red herrings crossed our paths because we thought only of the conventional.

SWATHING A NON-STARTER

It was naturally difficult to get away from the idea of wilting and maturing the crop. So many farmers decided that the way to go combining was to swath the crop with a binder, then, after some days of wilting, the green material could be easily removed and a perfect sample of grain obtained. Later, of course, we turned to the purpose-built swather, which we still use for certain seed crops, rape, linseed and so on. Alas the British climate made swathing a non-starter for cereals in next to no time.

Only if absolutely necessary should the weather risk to be taken on a swathed crop. One heavy thunderstorm between swathing and picking up with the combine was sufficient to cure most farmers of a misspent faith in our climate. This little flirtation with methods that destroyed the advantage of completeness came about partly from teething troubles, or if you like, through our stark ignorance of how to operate the combine in the United Kingdom.

The swathing idea was introduced because greenery would find its way into the sack, bag or tank. Often the grain could be reasonably dry and would otherwise take no harm, but the green weed seeds and foreign matter generated heat. Probably the common failing in early combine operators was one that is not completely unknown amongst users of today. This is the cardinal principle of always adding air blast, rather than reducing its force, when greenstuff is present in the sample going from straw walkers to sieves.

BLOWING BETTER THAN ROLLING

Now that most farms are equipped with sophisticated cleaning and drying machinery, users can get away with inefficient drum setting much easier. When grain flowed out of the rear of the combine with either straw or shavings, many operators reduced the wind, thinking that this saved grain. In fact, it meant that good grain was being rolled out with the chaff. In the opposite and correct case of applying maximum wind force, the foreign matter is lifted clear of the sieves and passes out. It happens that a certain amount of small grains will accompany it; the big difference is that these are the lesser grains. It is always better to blow out some grain than to roll it out with green material owing to too little wind.

Having decided that windrowing, or swathing, is a method to be used sparingly in our climate, it means that combines here have to be capable of clean cutting and good separation. The most standardised feature of combines, mowers, swathers, even binders, is that the reciprocating section and finger principle for cutting the crop has survived the test of time. Apart from foragers that employ flail cutters, the knife and finger cutting mechanism is standard on every make of machine.

In Britain we have to pay more attention to the cutting than in the Western Hemisphere. When heading the crop, a tattered and torn sort of stubble makes little difference, especially if, as in Australia, sheep are going to be turned in as scavengers. Here, and in most Continental countries, crops are likely to be lodged, or laid, apart from having some amount of green material growing with the straw. Despite our copious use of MCPA, plus modern and sophisticated sprays, much greenstuff survives to harvest in a wet year.

THE SERRATED KNIFE A BOON

One of the advances in recent times is the serrated edge knife. It is self-sharpening and much better than the plain knife. Over-serrated and under-serrated knives are available but I have found very little difference in the performance. Many manufacturers provide one serrated knife and one plain. The theory is that a plain knife, regularly sharpened, is much better for crops with a lot of weeds, undersown seeds, or rubbish in them.

This is more or less true but now getting somewhat outdated. Anyone who has spent endless hours sharpening knives to keep the old horse drawn mowers and binders going knows what a thankless chore that is. Moreover, few of today's farmworkers are very good at this task. Labour is too expensive to have a man sharpening plain knives. It is a better policy to keep a good stock of knife sections with serrated teeth and each morning put, a knife into the vice, hammer off damaged and badly worn sections and replace with new ones.

As many farmers want to haul the straw, this is another reason for wanting to have a good cut and a close one. For those who are going to burn the straw or lacerate for ploughing in, I never reckon it worthwhile shaving the ground, just to make a good looking stubble. A lot of extra strain is put on a combine that has to be driven with its cutterbar scraping the ground. Often this is inevitable, because of storm-damaged, badly-lodged crops, or if peas are being combined direct, but it should not be a habit.

When it is necessary to cut close, and when farmers turn to the plain knife and sharpening of yesteryear, this is often because the cutterbar sections are worn or damaged. No knife can cut cleanly unless it is reasonably tight fit, lying flat and running true. I attach some importance to the knife condition because the average farmer will continually walk behind the combine picking up chaff, shifting straw and feeling for blown out kernels of grain yet ignore the front end of the machine. However hard to believe, it could well be that 50 per cent of the losses are happening at the cutterbar end.

PICK-UP REEL FAVOURED

Were the crop not vulnerable at the cutting stage, there would be no need for the draper-type pick-up attachment. The ordinary finger-type pick-up attachment suffices for most windrowed crops, but the endless belt, or draper type, is much kinder to easily shattered seed crops. The bat-type reel has largely disappeared from our farms in favour of the pick-up reel. For a time we had a sort of combination reel of tines affixed to the bats, but now are almost standardised tine-type pick-up reels.

In the modern combine, the reel is adjustable fore and aft, up and down, besides being adjusted for speed of revolutions. The majority of such adjustments are hydraulically manipulated from the driver's seat. In the larger and most expensive combines, virtually all such adjustments are done from the operator's platform. In others of the middle range, the fore and aft movement is often a simple manual adjustment.

For UK conditions the pick-up reel was a great technical advance, yet it has been much abused by operators. Many combine drivers think that the reel must always be going full speed in its lowest position, but only in abnormally bad conditions is this necessary. A lot of the grain loss in combining is coming from a speeded-up reel threshing out grain before it reaches the concave. Equally, reels are often set much lower than need be. Some combine manufacturers recommend the opposite presentation of the crop to the cylinders, as was the case in threshing drum days. Butt-end first is the usual practice.

Like the cutter bar shaving the ground for no special reason, a fast running reel will only consume more power and cause more wear and tear. The reel would be running at the lowest revolutions per minute necessary to gather up the crop efficiently and present it to the front auger at the speed that the auger can deal with it. position, but only in abnormally bad conditions is this necessary. A fast revolving reel can half thresh the crop in over-ripe conditions. Nor is it any longer necessary to set the speed for the day. With today's finger-tip control, speeds and positions of reels can be continually adjusted in work.

BUTT-END-FIRST-FEED SYSTEM

The butt-end-first feed system encourages men to set the reel low. Indeed it is necessary. But there are times when the reel need hardly touch the crop; this applies to combining such pulses as beans. These days, peas are usually dessicated and combined direct, as against swathing in earlier days. Here the pick-up reel really comes into its own, but with beans it is the opposite. Usually they stand well and I rate them an easy combining crop.

It is still commonly recommended that beans be combined during early morning and late evening dew. Like a lot of folklore, this wants to be kept in perspective. True, this will mean less crop shedding, but I would rather shed a few beans than have a barn full of damp ones! With beans often the reel need do no more than run gently across the top of the crop, barely touching it. Beans are not very big yielders, compared with cereals, so forward speed may often be fast. Full speed full wind, slow reel and no sun, is the ideal recipe.

Beans and peas are so difficult to deal with in grain installations that are based on cereals, and especially if all auger-type feed is used, that the old-fashioned principle of getting the crop in from the field in storable condition is worth following. With cereals, we have rather changed our tune. In the early days of combining we aimed for a marketable sample and found it was difficult, though not impossible. Now most farmers take the attitude of rushing the crop from field to store, with speed the criterion and the state of the sample at a discount.

SEALED SILOS

Once in store, the grain can be cleaned, dried and graded under cover over a period of a 24-hour day if that be necessary to keep pace with the combines. At the other end of the scale, we now have many stock-keeping farmers on the sealed system of silos. In their case, directly opposite to the grain-selling farmer, they may prefer their grain coming in at 18 per cent to 24 per cent moisture content, the sample refinement being discounted. They just want to get the maximum yield.

While the grain-selling farmer also wants to reap as much as possible of what he has grown, he is concerned with low moisture both for keeping quality and for reducing the cost of drying. There is only one economical grain drier. The sun! With escalating oil prices, in fact, it may now be a good proposition to be over-combined by the economists' standards so as to reap more corn during the driest part of the day. For the farmer storing his grain into the following new year, it is certainly necessary to get the moisture down to 16 per cent and to be safe, 14 per cent.

AERATION DESIRABLE

Never forget that stored grain may take in moisture from the atmosphere during storage period. For long-term storage, some sort of aeration is desirable, and even necessary. For the malting trade, it is not only necessary to have barley free of foreign matter but also undamaged. Yet we are almost standardised on the rasp bar drum. The rubber-faced beater bars, as in the little Allis Chalmers combines that came here immediately after the second world war, seem to have given way to more uncompromising types.

The rasps may differ slightly in angle and depth between one manufacturer and another, but the basic function and the end result differs little. Despite the bludgeoning effect and the damage that can be caused, there is no need to do a lot of kernel splitting at the threshing stage. In the older combines, altering the concave was a laborious, painstaking and knuckle-dusting job. This is no longer the case and anyone who has a modern combine has little excuse for a lot of grain impairment.

Lowering or raising the concave, besides speeding up or slowing down the cylinder speed, plus the instrumentation that tells the operator what he is doing, is almost child's play. Where it used to be nuts, bolts and hammers, it is now a matter of a wheel to turn or a lever resembling an old brake lever on a motor car. The part that is not so easy is the judgment needed to turn a handle or slip a lever into the next recess on the quadrant. Many operators have just not got this knack, or sometimes operator comfort has made them so complacent that it has bred its own degree of laziness.

We refer these days to judgment and timeliness in cultivations rather more than in harvesting. One of the criteria for swathing crops has always been assessing correctly the moment to cut. But farmers have found desiccants far more effective than mere judgment with such crops as peas. The information is passed to expert fieldsmen in the case of quickfreeze peas where the decision to harvest is based on the computer-like adjudication of the tenderometer. With canning peas or stockfeed peas, however, the farmer may not have such sophisticated machines or expert advice on tap.

Waiting for such crops as peas, rape, or linseed to ripen 100 per cent is out of the question. Always, there has to be the inspired guesswork that will result in the optimum amount of crop gathered. With peas, one can reckon that the chemical desiccant has virtually superseded swathing. If one waits for complete ripening, the cutter-bar losses will be out of all proportion, so most farmers try striking the optimum by spraying a week or two ahead of harvesting date and getting most of the crop in the tank, most of it ripe and most of it undamaged. The tough strawed rape and the wiry strawed linseed may, or may not, stand the test of time in UK.

Whether combined direct or swathed, the vertical knife at one end of the cutterbar is essential for a good cutting job in such crops as rape and linseed. This assessory could be the difference between early failures and the more recent success with these oil-seed crops. This, allied to having finger-tip controls on the operator's platform on the modern harvester, has gone some way to alleviating the wretched performance that older machines put up. Even so, I would not want too large a percentage of my income depending on these two crops, nor too much of my capital invested in them and the machinery involved.

STRAW INHIBITORS

Straw inhibitors in cereals have partly achieved the same success for the combine harvester as desiccants have done in pulses. But their contribution is small by comparison. For one thing, many farmers do not use straw inhibitors, and secondly, they regard them as a waste of money. Light-land farmers treat such chemicals as a sort of joke, saying that they are pleased to get some straw to stand up to the cutterbar. They have no intention of feeding crops copiously with fertilisers only to inhibit growth in the later stages.

Again into focus comes the built-in judgment of the heavy-land farmer who has to guess the weather ahead of him when he sees bumper long-strawed crops in May and June. Sometimes a straw inhibitor chemical sprayed on can prevent or reduce lodging and thus materially assist combine harvesting. The number of cases where today's combine needs such chemical assistance are limited. Unless sprayed from the air, which is not cheap, wheel damage has to be taken into account with these late sprayings.

In the case of desiccants on peas, there is little doubt of the loss from wheel shedding of the crop being reimbursed in less troublesome harvesting. In the case of inhibitors on cereals, the case is not so good and two-halfpennies-for-a-penny returns are no good to bank managers as a general rule. Straw inhibitors can and do assist the combine in its task, but it is more likely that the botanists and geneticists at plant breeding institutions will wield the greater influence.

Perhaps the importance of the combine harvester as a gatherer of whatever result we have achieved from our year of labour in the cereal fields can be gauged from the fact that grain supplies more than 50 per cent of man's food energy. It is equally put into perspective when one recalls that something like 70 per cent of the land cropped in the world is devoted to cereals.

Looked at in that light, it could be that costs for combine harvesters of up to £30,000 are justified in 1977. Certainly the top-priced machines look to be on a limited market so far as farm and farmer size go. Add syndicate ownership and the contractors put a better face on it, but the money still looks rather much for three to four weeks' work a year.

Over 20 years ago (August 1956 to be exact) diagrams and a cut-away drawing of a new conception in combines appeared in the farming press. Inevitably it was another Ferguson idea. With a few modifications that looked rather more difficult to me than was obvious from the way the story was put across, it was claimed that the famed TE20 could become the powerhouse within a self-propelled combine. In actual fact, the principle was not lost sight of because we still have a similar type on our market in the form of the JF.

One of the surprising facts is that as costs have escalated, more farmers have not turned to the JF. What is illustrated by the design of the machine is the foresight that was inherent in that earlier concept. The cost of the power unit for a self-propelled combine then was only a fraction of what it is today, yet designers were obviously already worried. How right they were.

SELF-PROPULSION ALWAYS A WINNER

The British were absolutely and resolutely sold on self-propulsion in combine harvesters from the very beginning. It was not that we actually started on self-propelled models, but once a farmer had taken a good look at such a model, any salesman had difficulty in pushing the sale of either a trailed model or the wrap-round type using the farm tractor already available.

Massey-Harris-Ferguson (as it was then called) did not market their little brain-child, in their wisdom. At that time anything other than a self-propelled unit was to be doomed to a sales failure.

One cannot help wondering whether farmers will ever take a second look at the wrap-round combine or the trailed types. The reasoning behind looking at trailed types would be that tractors now have such tremendous horsepower available at the pto which they did not have in the early days of combines.

Thinking of the early post-war years, when the International 62's, the Allis-Chalmers 60's and those 13 dull-brown Claas models arrived in 1947, most farms were muddling through on standard Fordson tractors and the TE20's. Already, a year earlier, farmers had noticed a few Massey-Harris 21's which had arrived from the prairies, with all their limitations in this country.

Despite the fact that International and Allis-Chalmers models had their own engines, and consequently the tractor did not have to be all that powerful, this type was not to last. Even the Allis, sporting the little Allis B tractor engine and with several B tractors around the countryside to support it, to say nothing of its rubber-faced beater bars, struggled for recognition. Farmers did not want to know.

The trailed Internationals with enormous cutterbars of at least three metres, and the new arrivals from Germany with a startling 2.14 metres, were to be seen in reasonable numbers. The MM G8, the Oliver and others had already looked in on our market. Leon Claey, over in Belgium, obviously had the entrepreneurial skill, plus an in-built market research, that deciphered the British temperament far better than our friends from the new world. Leon Claey (now Sperry New Holland) set out to build self propelled only. His judgment turned out to be a classic in the history of farm machinery in Europe. Like Henry Ford and Harry Ferguson, Leon Claey's name will always find a place in the story of farm machinery in general, combine harvesting in particular.

The sales success story in combine harvesters in the UK market belongs to Claas from Westphalia, Germany, although the Claysons must have matched it in recent years.

Claas make no bones about their direct approach. Therein lay combining principles that students and beginners would do well to

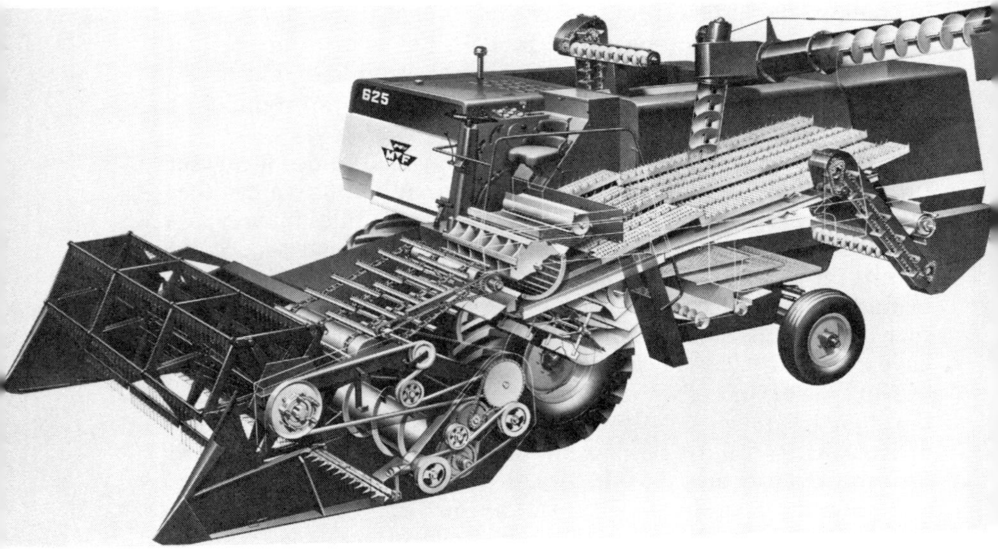

46 Cut-away picture of the MF 625 combine showing all the basic principles of combine harvester design.

47 Sperry New Holland Model 8080 giant combine harvester in action in wheat. This is a pre-production machine not fitted with a cab.

48 The Claas Dominator 100 imported to the UK by J. Mann & Son Ltd; Saxham, Suffolk.

49 The MF760 combine is the biggest ever marketed by Massey Ferguson in the UK. One of the 'new generation' machines in the trend towards higher productivity by providing the operator with a controlled, super-comfort environment, it has a luxury cab pressurised to keep it dust-free, with full air-conditioning providing refrigerated or heated air as required. It is offered with hydrostatic transmission with 5.48 metre or 6.1 metre 'Quick Attach' tables. Latest addition is the 'Power-Flow' table which increases the distance between knife and auger to 500 mm.

50 The Danish JF combine built round a David Brown tractor. Most of the well-known tractor makes are suitable and the tractor is removed for normal usage during the rest of the year.

51 The John Deere 965H Sidehill Combine will work on an incline of 20 degrees, with the cutting platform parallel to the ground and the operator platform on the level.

note. First, gather the crop efficiently at the front end. So, say the Claas designers, supply grain lifters as standard. I am not a fan of lifters, because they have too much of the old reaper look about them. They are useful in peas and any laid crops, but not the be-all and end-all of gathering a damaged corn field.

The large distance from fingers to auger in the Claas helps to ensure that the crop is not pulled out by its roots. This is important in that the dirt accompanying pulled-out straw can cause a lot of threshing inefficiency at concave and sieves. With the British sold on self-propelled combines because of our difficult climate and the consequent weather-damaged crops, the often overlooked advantage over the trailed combine was the ease of forward and aft movement. The pusher combine is how many farmworkers described the self-propelled ones in the early days. How right their terminology was!

It was the action of pushing forward into the crop, then reversing and having another go that mattered. Above all else was the ability to cut a field 'one-way'. Opposite to the binder, we had got this job of re-educating ourselves to take the crop laying back from the table, not towards the direction of travel. We could back around the field anywhere any angle that suited us. This is what made the self-propelled combine a real 'hit' with the British farmer despite the enormous cost increase. Conscious of this, Claas now offer Retromat, which enables change of direction from forward to reverse and back again.

With Retromat transmission it is only necessary to push the clutch pedal to the floor and release it; change of direction is then automatic. It is particularly useful in adverse conditions because after reversing, the operator gives another push on his clutch pedal, which takes him into forward motion again at the original pre-set speed. One word of warning: changing from forward to reverse suddenly can spell danger for anyone or anything at the rear. Many farmers have a habit of following up behind unknown to the driver, mainly to see what grain is going out with the straw or the cavings. A sudden change of direction without turning the head could slightly reduce the number of bosses in the corn fields of England! This is always an inherent danger with the torque converter principle. I have personal knowledge of at least one tragedy with a torque-converter-fitted tractor reversing over an unsuspecting human.

One of the world's largest combines in commercial production is the Massey Ferguson 760. This is a nine-tonne monster that has a particular emphasis on servicing in that the front-end mechanism can even be reversed to clear blockages under power. One big difference between Claas and Massey Ferguson combines since the MF 500 series was that the MF rotary screen that took the dust and dirt away from the driver was in direct contrast to the exaggerated auger to finger distance on the Claas that allowed so much dust to rise.

Now, thankfully, all these problems have been solved. Farmers can now buy combines strictly on merit, because with a cab fitted dust is not a major factor.

One MF 760 has been known to harvest 100 acres in a day. In just one hour, a 760 can harvest enough wheat to keep an average family of four in bread for 60 years. The low silhouette of the Massey Ferguson combines that came with the 500 series in the early 1960s, plus the cleanliness factor that gave the operator a long overdue environment that would not give him farmer's lung disease, made many of us wonder what happened to them in the late 1940s. Perhaps their design innovations culminate in the unique paddle crop elevator at the front end of the 760. Equally inventive and just as good was the adoption of automatic table height control.

AUTOMATIC HEIGHT CONTROL

After others had nibbled at the edges of a possible market for automatic table height control, using under-table sensors, there is also the overriding manual control in the driver's platform. There are times inevitably, mainly in bad conditions, when the response is not fast enough under automation. Nor can it be good for the mechanism if, in soft conditions, there is going to be a lot of machine sinkage and the cutter bar jumping up and down all the time.

One good bonus with automatic height control is that is should prevent bulldozing. Many of the older combines had a tendency to bulldoze into the ground in lodged corn and in soft conditions. The early Claysons had this fault and, for that matter, so did most combines. It was not difficult to get the fingers of the popular Massey Ferguson 726 and 780 models underground. The surprise about automatic table height control is not that it eventually arrived, but that it is not more common.

GRAIN MONITORS

A little overdue, but not accepted as the ultimate in the combine design, are grain monitors. Such gadgets virtually reduce the operators function of observing the cutter bar and watching the monitor

52 The new-type flat, dry-type sensor used in the RDS Mk 3 Monitor. In this case the combine is a Massey Ferguson 760.

53 Diagram showing the positions of the three full-width sensors of a new, dry pattern fitted in the new RDS Mk 4 Monitor. The meter unit in the driver's cab can indicate the scale of losses at any individual sensor, or can present combined readings. The introduction of the third sensor to detect airborne losses makes it possible to use the instrument to achieve accurate sieve and fan adjustment.

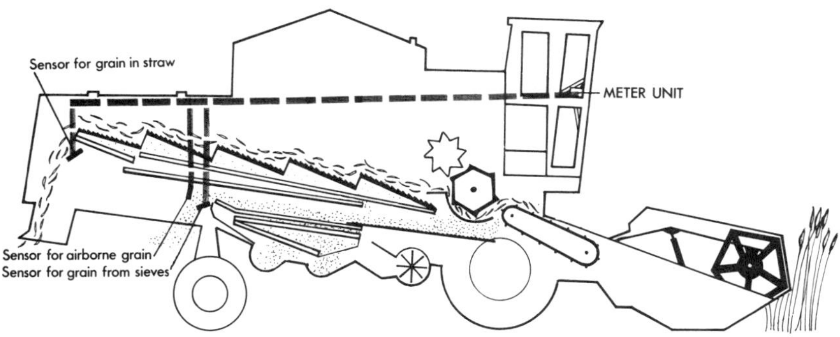

A grain sensor consists of two sensors which extend to the width of the combine, one behind the straw walkers and one behind the sieves. They record the sound of grains falling on them and convert the result to a quantitative meter reading on a dial in front of the operator.

The man on the seat is then supposed to adjust his speed of travel to maintain a selected meter reading. It is claimed, perhaps a little optimistically, that speed will always be the optimum for crop density, weed content, moisture content and ambient humidity. Maybe speed will also be appropriate to the field gradient, but more about that later. More, too, about the drawbacks, but I do not wish to be churlish about these necessary aids to modern farm machine usage. The Mann-Claas partnership now have the Mann-RDA monitor in three-sensor configuration.

It is claimed that three sensors instead of two is an aid to setting up the combine for work, as against merely being an aid while driving. My reservation that led me to mention possible drawbacks is that there are the two extremes of operator performance, as against machine performance. The modern combine is so advanced in technology that an operator may be failing to realise the machine's full potential. The monitor could bring such a driver up to the mark. Literally so, of course, if he can be instructed and persuaded to drive to the mark on the monitor in front of him. This does nothing to decry either the theory or the practical aspect of monitors, automatic height control mechanism, or the crop density allied to forward travel control that will come later. I only want to reiterate my obvious science in farming. The weather factor alone will keep it that way.

In easy conditions, a top-class combine operator can often be getting 5 per cent more grain per acre than an unskilled driver. In adverse conditions, the margin between the good and the incompetent operator can rise to as much as 15 per cent. In some ways, the farmer's cry for automation is no more than a plea from the heart to surmount operator inefficiency in machine usage. Hence, in combining, as in everything else, we alleviate risk so far as is possible. The primary threshing cylinder, placed ahead of the main threshing drum, was this sort of idea.

If 20 per cent to 25 per cent of the grain could be got out of the straw before the main cylinder did its job, so much the better. At least, that was the theory. I am not really sure that it ever worked! Certainly the work of the straw walkers can be reduced and the work of the sieves evened out a little. The less work the walkers do, the better it is, but there is no way of avoiding the sieves except by throwing it out at the back end. Massey Ferguson have even tried to sort out that little problem with Multi-Flow, an assessory that gets what grain it can from the straw after it has travelled the length of the walkers.

RE-THRESHERS

Re-threshers are another aid to getting all the grain, as we know from their use by Massey Ferguson and Clayson. These little gadgets have been around on the Massey Ferguson 500 series for a long time now, re-threshing the tailings or screenings. They are not to be discounted, though stones that jam in them when it is necessary to scrape the ground in badly lodged crops can cause a lot of driver movement from seat to re-thresher.

Several years back, Allis-Chalmers, who, like Ransomes, are no longer in the combine province, brought their Hillside model to England. I mentioned earlier that the grain monitor may have some relevance to land gradient, it is well known that a lot of grain can be lost on hillsides. For a long time manufacturers rather foolishly stuck to sieves as flat as the land they expected cereals to be grown on.

Nowadays, most combine makers put dividers into their sieves so that all the grain on the sieve at any given moment cannot shift to one side and spill out either at the rear or at the side, or both. Not surprisingly, because of its price, Allis-Chalmers found little interest in the Hillside, despite some sort of foundation having been laid down for their name in the combine sphere by the little 60, the Gleaner and the 5000.

COMBINING HILLY AREAS

I am by no means certain that a sufficient change of emphasis has not come about to justify a second look at the problem of combining in hilly areas. John Deere think likewise and have a model called the Sidehill. The John Deere 965H has been developed over in Zweibruken, West Germany. John Deere established themselves in Germany some years back, when they took over the former Lanz establishment. When announced, the 965H was less that 20 per cent more expensive than its equivalent. This is still fairly big money on a modern combine, but there could be areas in the UK where that percentage of price inflation is justified. The combine remains upright on inclines of up to 20 per cent or 11·5 degrees.

The interesting thing is whether the self-levelling combine, albeit very much more expensive, becomes common here, or for that matter in any reasonably flat territory. As with the pea viner, if it means a lot more corn in the barn rather than on the ground, it is bound to happen.

All sorts of figures are conjured up as to the amount of waste that there is in cereals, potatoes, sugarbeet, or any other crop on UK farms. One cannot imagine that such figures are anything more than a good guess. It all comes back to common sense and such aids as the grain monitor.

As far back as 1964 a youthful graduate of the Saskatchewan College of Agriculture in Canada tried out a device for automatic speed control on combine harvesters of the self-propelled type. With it, it was said, a farmer could pre-determine how much grain loss he would tolerate. Forward speed was regulated according to the density of swath going in at the cylinder.

ROTO-THRESHING—A CANADIAN CONCEPT

In the same year other Canadian farmers, upholding their reputation for ingenuity, went for what was called roto-threshing. Rather like a drum-type beet cleaner, the Canadian idea was a drum revolving at about 35 rpm.

The concept was simply a revolving drum replacing sieves and walkers, surrounded by a shroud. Fans created an air blast which passed between shroud and drum and separated grain and chaff, while the straw worked its way along the length of the drum.

The Sperry New Holland TR70 did away with the cylinder as well. Its two auger-like cylindrical rotors run lengthwise and contra-rotate and are surrounded by 'threshing grates'. Straight from the intake elevator, the crop is drawn between the rotors which move it along and against the surrounding threshing grates. Grain is then discharged by centrifugal force onto a grain pen, and then onto sieves in a more or less normal way.

Our own NIAE also added its contribution to replacing the conventional with the Air-Stream combine. In other words, and this really applies to the others too, emphasis was put on air as the separator.

As we know, air blast is very important, and the NIAE idea of all-air separation, rather than air-assisting mechanical separation, has about the same chance of turning out to be the system of the future as anything else. An evaluation programme by Sperry New Holland has been going on in North America, and the initiation of the TR70, with the emphasis on soya beans and maize in a climate that is hardly similar to the UK, can be taken as an indication that straw walkers are here for a few more years.

Incidentally, mention of tell-tale rows of self-sown corn does bring up the same tell-tale lines that often illustrate where the dividers went. Whether of the torpedo type, or what looked like a piece of bent wire on some of the early Massey Ferguson combines, dividers do manage to shell out a certain amount of grain. More, in my view, than is realised and more than any manufacturer would want to admit.

As I have indicated, the treacherous-looking vertical knife is now offered for such crops as oilseed rape. Maybe we could do with it for grain too, but I hope not. Not if good, effective dividers of simple construction can do a better job than at present.

Sperry New Holland as well as the Clayson 8080 favour a four-speed gear box coupled to hydrostatic final drive. As with Retromat on the Claas, reverse gear comes via the hydrostatic unit without de-clutching and at half forward speed. So far as the British market goes, the 8080 aligns itself alongside the giants from Massey Ferguson, Claas, Dania and Fahr.

Conscious of the hillside factors that I have mentioned, a point has been made of the excellent weight distribution on the 8080. This combine gets a rating of 78 per cent weight on the front axle and 22 per cent on the rear, but with a five-tonne tank full of grain, the weight change is only of the order of 2 per cent, at 80 per cent and 20 per cent.

As in the tractor sphere, hydrostatics are the inevitable end of the combine transmission story. Assuming ourselves to be completely and irrevocably addicted to the self-propelled machine, the instant torque-converter action of back and forward movement in these sort of machines is a must.

SIZE AND SHAPE OF CABS

One very practical, seemingly unimportant and even criticised fact of combine harvesters that I want to mention is the size and shape of the cabs that are quickly becoming a standard feature.

For some psychological reason best known only to themselves, many combine drivers seem to think that they should guide the machine round the field in a standing position. This could go back to the old iron-seated tractors, where it was the only comfortable thing to do, especially if one had lost his sack! According to the individual's reactions and according to how relevant the fatigue factor is, I reckon that there can be a five to 10 seconds lapse when a sudden calamity arises before the foot-operated clutch gets depressed. In those seconds, the driver, I notice from years of observation, usually

54 Wilder hydraulic spreader kit, here fitted to a Wilder 1.83 metre straw chopper. The kit comprises two hydraulic motors driving two 80 rpm counter-rotating spreader plates. Power to the hydraulic motors is via the tractor hydraulics.

55 The cab for the International combines provides excellent all-round vision and can be equipped with an air pressurisation unit to exclude dust. A dust extractor fan can also be obtained to pick up dust at the feed conveyor and direct it to the ground.

resumes his seat first, depresses the clutch pedal second and then sets off to remedy the trouble.

As with tractors, a farmer considering which make of combine to buy will more than likely go for the one with convenience of service. In fact, it is fair to say that service in the harvest field is to most arable farmers the most important aspect of farm machinery service.

OUTSTANDING COMBINE HARVESTER

I have little hestitation in giving my own opinion on the merits of combine harvesters. I regard the Clayson as one of the outstanding combine harvesters in the world, and I also place this firm as being in the forefront of farm machinery development in this country.

The first self-propelled combine that I remember in Britain was the International 123. Disappointing though it may be, when I saw it at work in Norfolk during the war years, I thought it was to be the forerunner of many, especially with several of the trailed Internationals already here. But the arrival of the Massey Harris 21 seemed to change the course of events for no particular reason. Soon the Masseys were to dominate the combine scene until Claas came in with their already-mentioned phenomonal success.

Way back in 1969, NAAS (now ADAS) had so sensibly looked at the underrated factor of time out of work. The Claysons won on a 2 per cent figure. This was really written confirmation of what farmers had gleaned from observation. It was a remarkable combination of ruggedness and reliability in an exacting job that eventually took Clayson to the top of the sales charts.

I hope that combines go automatic in operation so far as is possible, rather than gimmicky. In fact, any farmer should try to avoid gimmickry in all farm machinery.

First Ransomes, then Sperry New Holland with what looks like a miniature peg-drum, have tried the double drum threshing, while John Deere have their Cross-Shaker fixed across and above the straw walkers. Then Claas introduced a similar straw delay device, if that is the way to refer to it. Massey Ferguson, with rather more science, had long since tried multi-flow, referred to earlier.

All these innovations are simply trying to separate grain from straw. All are assisting drum walkers and sieves in the job they were designed to do and did in the stationary thresher. Provided all these developments either get a better record of grain loss, or increase output by, say, 5 per cent or 10 per cent, all well and good. But I do not want to see farmers having to pay for additional engineering and metal to put right what should have been accomplished by the basic machine layout.

I remember spending some hours, and achieving the inevitable skinless knuckles at the end of the operation, putting risers in a Massey Ferguson combine. Then we went back to work. Then we took them all out again.

ROUND-AND-ROUND WORKING BEST

There may be ancillary reasons of travelling up and down, such as when there is a stationery trailer on the headland, or even to suit subsequent straw baling. Combining in 'lands', sometimes going the length of the field for a piece a fraction of the cutterbar width, is ridiculous in the extreme. Up and down, where the unloader spout will often be pointing inwards to the crop, assuming continuous emptying is being practised, is equally silly if manoeuvring has to take place for emptying. With round-and-round working, the odd passage can be cut across big or awkward fields for the trailers, and my experience is that this wins handsomely.

Over the years various methods of cutting off odd corners and going back at the end to do them, thus rounding the field corners, have been tried. It is not to be recommended except by those who like a complicated life. Most drivers known that a loop at the corners is much quicker than gear changing to shunt round, and the modern combine with its finger-tip hydraulic controls can be much speeded up for the loop. Individual brakes can be brought into play too.

Back in the 1950s we tried the German idea of low density baling by a press attached to the combine. It failed because there is no time for complication in harvesting methods in an English summer. My experience was that the press was more trouble than the rest of the machine. The only way to break even on straw sales is to make the purchaser pay in advance and sign an agreement to clear the land. Even muck for straw is not foolproof. I know, I have tried most of the straw gimmicks!

No manufacturer seems to have made a fortune out of attached straw choppers like the Rekord. Tractor-drawn choppers, such as Ugerlose, or spreaders like the Wilder, had some popularity, but not over much. Also, no manufacturer has made a big hit in straw spreader sales. The only straw spreader that I tried and got to work properly was the one for the side-discharge Allis-Chalmers 60 combine in the late 1940s.

The truth is that not too much research has been warranted, because farmers, despite warnings of a humus content drop in the soil, have not shown any great interest. They have long since found that a long stubble puts back some humus, and it only seems to be NP & K that produce the crop anyway. The ups and downs of the humus content is a slow process. Not all farmers want to live as though they are going to die tomorrow and farm as though they are going to live for ever!

There is little difference now between our four main market contenders—Clayson, Claas, Massey Ferguson and John Deere. While the gap gradually closes, I still rate the Claysons as supreme at almost every level. After the Claysons, I turn to the John Deere, then Claas, Massey Ferguson, International and Laverda. There are other good machines on our market such as Dania, Fahr, Fortshritt and the UK-made Lely Victory from Fisher Humphries, but I feel that a network of on-the-spot agencies to give instant service has to be an overriding consideration.

The Victory can boast of a feature above all others—that of the two-section, fold-up cutterbar. For contractors with narrow lanes to negotiate, this has to be a consideration, though a point about all the best combines now is the easily detachable table and the availability, as an accessory, of a table trailer. In this section above most others, the importance of sound service cannot be reiterated too often.

Chapter 6

THE STRAW HARVEST

THROUGHOUT the 1960s grain-producing farms appeared to await the coming of some Messiah who would tell them how to deal with the waste product of corn production: namely, straw. For the purely arable farmer, without stock or a market for straw, a box of matches is probably the only solution. Certainly it is more economic and less harmful than the average urban dweller would believe.

The early 1970s saw a more positive approach to straw utilisation only because urbanised people and the surfeit of do-gooders had created such a fuss about burning and, more to the point, cereal prices had hardened somewhat. The backlash from much advanced grain prices was that feedingstuffs became expensive. Thus livestock farmers were clutching at straws, as it were, to reduce the price of producing milk and beef. Poultry and pig producers were in the same position, even worse. Ruminants were the obvious target for a better utilisation of straw.

The match box had no trade-in value when the farmer went shopping for a baler. The market was buoyant and sales were good. The neat, tidy and tight, rectanglar bale was a foregone conclusion if this wretched straw had to be gathered.

In came the machinery people with the big baler and the half-tonne bale. With hay, we may be talking in terms of three-quarters of a tonne, but half a tonne of anything to be picked up and carted somewhere, with our archaic transportation in the countryside, was a problem to be reckoned with. Of course, farmers did not have to buy big balers, but they did.

ROUND BALES

Many years previously we had tried round bales with the Allis-Chalmers Roto-Baler. These were no good, or so it was said, because the string was only a wrap-round. They were difficult to stack and, in any case, no farmer could afford the time for the tractor driver to have to put his foot on the clutch and stand still while the Allis disgorged its bale. We are now back to where we came in! That is, excepting for the fact that the bales are now many times larger, the tractor still has to come to a halt to eject, and there is the problem of how to deal with these newcomers.

Farm machinery has always had some resemblance to fashion in women's hats, in that things come and go, only to return. Jones had a self-propelled baler in the UK before most farmers knew what a baler looked like. The loose straw collector was still in vogue. Ransomes were distributing the German Lorant baler. This was a low density baler that looked at home either out in the field or baling behind a stationary drum. Not only did the Lorant do a good job in its day, but it survived the challenge of the high density baler on many farms, until such time as we found out how to turn at the corners without smashing the power-take-off drive at some point!

The late 1940s and the 1950s saw all types jockeying for position. The old rear-towed hay and staw gatherer, the Jones Invicta self-propelled baler, the Lorant, the Internationals that went into production here as early as 1950 and, of course, the baler to end all baling, the Raussendorf Press attached to the combine. The baler won hands down. International Harvester were one of the first in the field with power shafts anyway, so it was no surprise that if the baler had to be pto driven, they would be in the forefront.

LOW-DENSITY BALERS

It was surprising how well such types, as epitomised by the Lorant and the Welgar low-density baler, stood their ground here and on the Continent in the face of the high density challenge. Some manufacturers adopted the engine-driven baler, with the tractor used only for draught. As in the case of the Allis-Chalmers '60' combine and the International '62' combine, farmers did not tolerate this complication for very long. I had used an engine-driven baler with fair results, but many operators were known to spend many hours cranking engines.

The ram baler was here to stay, but the engine was for burning. When I traded in my combine with straw-tying press affixed, the machinery dealer asked if I would mind backing the combine up to the old horse pond that was a few yards away. When I queried this strange request he said that he would come back with his spanners and drop the press into the pond!

By 1966 we had 87,000 balers in England and Wales. Sadly, the Jones Invicta was a thing of the past, and there has been no replacement in the sphere of self-propulsion in the field of combine harvesters and more recently in the province of forage equipment. The Lorants have gradually faded and the Allis Roto-Baler has long since become a museum piece.

The round bales were a nightmare when it came to stacking, though we are inclined to say that as though there are no problems stacking oblong bales. It is like building with bricks and no mortar. The safest way not to have a collapsed stack seemed to be the method of always trying the corner bales to each other, so that about four bales are involved. Either that, or two corner bales tied across, near the stack middle. Gable roofs, as with sheaves, were sometimes all right and just as often all wrong. Water would find a way in, or the wind would play havoc. So often the roof went flat, with weather protection from plastic or tarpaulin.

PERMANENT COVER FOR STRAW

Straw is normally for use on UK farms either as fodder, in the case of good barley or oat straw, or as litter if wheat straw, plus, of course, the low grade barley, oat and bean straw. Pea straw can often be useful fodder, but rape straw and linseed straw, unless required as flax, are often more trouble than their worth. Despite all the efforts and the many pet ideas for keeping stacks of straw watertight, in the absence of a Dutch barn, it is gradually becoming obvious that if we are now to look at straw as a marketable crop, then permanent cover is imperative.

Fortunately, many farms that lack a Dutch barn have found a new use for buildings such as old-fashioned bullock yards that have become more redundant. The important point is that we could be on the verge of taking a second look at straw. This would mean straw for sale, as against straw gathered for the farm livestock. In any case, with a slight trend to organic irrigation from dung pits below slatted floors, livestock farmers are often using less straw than at any time in the past. I am not a fanatic about this departure, because if often seems that a lot of additional machinery and work is involved, all to avoid the use of straw.

Yet copious straw is available, almost gratis, in the arable east of the country! We witnessed the big straw lift of 1975, when thousands of tons were shifted from the east to the west. It proved that this is both a difficult operation physically and no less worrying financially. Straw is expensive once baled, moved and unloaded hundreds of miles away. For the arable farmer who sells, then finds that he is lumbered with wet straw to burn, or expensive bales that he has no use for, it is not a proposition.

If the great straw maelstrom of 1975 did nothing else, it did, through the auspices of the NFU in many cases, expose to arable and livestock farmers each other's case. It proved my point, that straw is likely to be a loser to all who dabble with it, except, perhaps, the man with the match box.

The main factor in the straw-burning story is that the arable farmer can get on with his autumn work immediately. That is important, beyond all doubt. Burnt land cultivates better, ploughs better and some weed seeds may be destroyed, while other seeds are encouraged to grow and get turned under. Some nitrogen is lost by burning but some potash and a little phosphate is left in the carbon residue.

BURNING—THE ECONOMICAL WAY OUT

Despite local bye-laws, I doubt whether burning is going to be completely stopped. It is so often the only economical way out. It is as much an act of husbandry as the burning of sugar cane in Australia and South Africa. The outcry against burning that the farmer has had to tolerate in the past few years comes as much from ignorance of the true husbandry facts as from anything else.

Following the increase in feedingstuffs prices that had in turn followed cereal prices, our initiation into utilising straw on a rather better commercial footing came, fittingly enough, in the Eastern Counties. The BOCM-Silcock animal food company had spent £300,000 on a pilot plant at Kimbolton, Cambridgeshire. Two further plants, in Suffolk and Yorkshire, took expenditure over the million pound mark. Now, more millions are going into plants, so that there are now (1977) eight in operation and a processing capacity of 200,000 tonnes of straw.

Converting straw into animal food will hardly cause any great threat to the soya-bean producers of the world, but in a wasteful age it will help to produce some more food from our own resources, besides helping our balance of payments problems. In a small way, claims are almost automatically made that by preventing straw burning by offering remunerative alternative uses, pollution of the environment is reduced. I hope that card does not get over-played. An economic outlet for straw is the arable farmer's main concern.

To reduce bulk for long journeys, machines for re-baling have come onto the market. In other words, high, high-density baling. We have gone from low density to high density, we have tried wire but settle mainly for twine-tying and now find it worthwhile to double-bale. Having got the straw compacted as much as is possible in the field, it is possible to re-ram and re-tie the bale, thus cutting costs when transporting over long distances.

WHOLE-CROP HARVESTING

I have chosen this chapter, rather than the combine harvester section, to mention a possibility that should not be overlooked. This is the idea of whole-crop harvesting. I have seen it done with oats and barley in Ireland and with maize in America. I have no doubt that whole-crop harvesting is being practised in many lands with several different crops and certainly we, in the UK, are carrying out experiments. This sort of second look at the fundamentals of the crop-machinery relationship has to be done occasionally.

Having discarded the combine press, it seems more logical to think in terms of adding some type of cutting mechanism to the baler. The logic of separate harvesting techniques for grain and straw, only for the livestock man to mix it all up again, is being seriously questioned. I am not surprised. It is not beyond the realms of possibility that whole-crop harvesting should be followed by a separation of grain from the straw. Proprionic acid could be brought into use for preservation if necessary, but I would rather doubt the economics of merely changing the venue of separation from field to barn.

For the livestock man, whole-crop harvesting, which would virtually mean the end of the combine for him, must have some attractions. Alternatively, it can be the combine all-crop machine, to the exclusion of the baler, but one suspects that the pressing and tying operation would be essential for transport. In this enlightened age, I am surprised that more livestock farmers have not looked at whole-crop harvesting, when all the technological and scientific knowledge available to us could so easily prescribe the additives for a ruminants ration, after due analysis of the gathered—and dried if necessary—whole crop.

Whole-crop harvesting could just as easily be accomplished by a modern forager, but so much depends on whether we consider the fully ripened crop best or whether we prefer grain in the 'milky' state. Every stockman knows that whole grains passing through the guts of his animal are rather wasteful, unless bullocks and pigs are run together with the latter as scavengers. Not only did our grandfathers do just that, but I have seen it done in the Western Hemisphere with whole-crop maize.

To envisage some sort of whole-crop harvesting in this country is to think in terms of a grinding, milling, or processing operation after the field harvesting and before the final feeding. Grains going through the innards of first the ruminant and then the pig might not make sense. But the logic of whole-crop harvesting cannot be denied, if we divide the purely arable farm from the farm where all field produce is fed.

RAM BALER

For the time being, we rely on the ram baler, where the reciprocating ram or plunger compress the hay or straw into the bale chamber. The retracting time pick-up assembly at the front of the baler is more or less standardised. The cross-feed method varies between auger, sweep fork, tine bar and moving carriage, but the end result is two bands of twine or wire (very occasionally) round a rectangular bale. Knotters can hardly vary very much because the end result has to be identical. Unlike the binders of yesteryear, there are no longer big variations in knotter performance.

While nobody can guarantee the modern knotter against failure, it often gets blamed. Failure to select the appropriate bale density for the crop being picked up can put a strain on the whole twine-tying mechanism that was not intended. Excessive density can mean that the twine gets pulled out of the retainer, while too little density may result in the bale being too loose for a clean pull off the knotter bills.

Bent bales with one string tight and one loose are still a common occurrence. Usually they come from the throw of the feed tines being wrong, a blunt shear knife or ledger plate, and incorrect clearance between knife and plate, as well as starting too early in the mornings when the swath may be dry on top and wet underneath. It is surprising how many operators can invent their own route for the twine. Modern baler operators have not always had the experience of the horseman who has threaded the needle of a binder a million times!

The tension of the twine from box to knotter is something that should not be overlooked. A few handfuls of loose straws that manage to wrap themselves round the twine en route will be tolerated rather better in some balers than others. They will be noticed and cleared by some operators better than others!

Very few knife sharpeners keep a broad enough surface on the knife edge. The idea seems to be to round the sharpened area rather than keep it flat or to put an edge on like a razor blade, which will go off quickly. I am talking of all knives, rather than twine knives, shear knives, mower knives or plain combine knives. Knife sharpening is like taking a watch to pieces. First thing to do is to look.

PICKING UP BEHIND BIG COMBINE

Since the early days of baling with, say, the International 45, tractor drivers have learned a few things about sharp turns with balers. As it happens, the hydraulic operation of the pto has spared us a few grey hairs, while watching a man negotiate a corner. Still it is a good thing not to be too complacent about corners, though it is no longer difficult to deal with quite sharp bends. One noticeable difficulty nowadays is picking up all the swath behind one of the giant combines. While the fitting of, for instance, a Vicon finger wheel rake on the front of a tractor can put two swaths into one, if that is desirable, we do not want the expense as a standard requirement.

While some manufacturers guard against a swath too wide for easy gathering, it is not difficult to cope with the problem at the combine end anyway. A local backsmith or the farm mechanic can easily weld a few strips of metal together to make a deflector plate, which can be bolted at the rear of the straw walkers. For straw to be burnt, the deflector is better not to be in position. At the moment of writing (1977) the widest pick-up of any baler on the UK market is the 1.73 metres of the Massey Ferguson 128.

Most balers will now deal with sisal or plastic twine, with self-compensating features to cope with variations in the twine. Another feature is the ability to apply tension to the bale by virtue of hydraulic operation from the comfort of the driver's seat. This must not be overrated, because it is difficult to judge the weight and tension of bales without getting off the tractor. Most bales are mechanically handled now, but if not, it is always a good piece of staff management to see that the man who did the baling is in the team collecting the bales! Just in case he did not leave his seat and check bale weight and tension in the chamber often enough, he should be taught a salutary lesson about the muscle power, or, put another way, the science of kinetics!

DEALING WITH BADLY LODGED CROP

Everyone who has ever done any baling knows that one great difficulty is in the badly lodged crop, where the combine driver has had to do a lot of shunting back and forth. That way we get a lot of lumps in the swath, which used almost to wreck the old-stagers and which do not tend to improve the mechanics of even the most modern of balers. While the modern balers deal with these large heaps much better, it is still often necessary to de-clutch and ease forward into the heap. The man using the baler is not blessed with the advantages of self propulsion, as his colleague on the combine almost certainly is. This man is still in what we call the neck-craning era, and if he forgets to turn his head and ease into the odd heap that he sees going past his tractor fender, he may have to spend several minutes out of his comfort cab to clear the blockage.

WAFERING

Not surprisingly, farmers and farmworkers grew tired of throwing bales around by hand as the more sedentary age came upon us. A few blocks, some mis-tying and many shear pins to replace out in the field, was nothing compared with keeping stacks upright, or even getting the bales from field to barn. Many times I said and wrote that we had to go big or small. I thought it would be to the small end with wafers, Massey Ferguson having brought the Waferer to the 1961 Smithfield show.

We know the difficulties that affect wafering, or pelleting, with our climate. What works in California may not work here. Given the choice of big bales to handle mechanically, because that is the only way to stop the drudgery, one would gamble that the farmer would go for a large version of what he already had. If bales went small, they could be handled pneumatically but if they went big it would be mechanically. I am by no means convinced that the big baler is here for evermore.

The farmer lost his option on whether to haul the bales by hand pitching or mechanical pick-up. In 1972, Howard came out with their big bale, but they were more or less the shape of what had become the traditional. They were square to oblong with sharp corners and made for stacking, to say nothing of picking up with anything from a buckrake upwards. The big baler was here, an honest-to-goodness package.

BIG ROUND BALES

By the time International had introduced their big round baler, we were back to the melting pot but convinced that bigness was right. Bales had to be big enough for mechanical handling without any reservations, but suddenly few people knew what shape they wanted. The men could not stack the round bales that the Allis-Chalmers made, so there was no hope for the round ones. How wrong! Within a year or two the big balers had got 10 per cent of the market, mainly because the bales could be picked, hauled and stacked in about one-third of the time. What is more, most of the bales were round, not square.

At the 1976 Smithfield Show, Howards up-dated their big baler. Each bale is equivalent to about twenty-three ordinary bales of the small type. Handling of the bales is by a hydraulically-operated gripper which can be attached to most of the heavy-duty front loaders. It is a sign of the times that Howards have strengthened the transmission to allow tractors of up to 90 hp to operate it. Such a machine is capable of up to 2.40 hectares per hour.

BALER WITH A DIFFERENCE

More important, so far as trends in big balers go, was the fact of the same group of companies announcing a new round baler at the same time. The Claas Rollant is the baler with a difference. Steel rollers, numbering twenty-one and easily removed if necessary, have been made to produce a bale with a high outside density for weather-proofing so far as it is possible and a soft core to aid barn drying with hay. It is a fixed dimension bale 1.83 m in diameter by 1.52 m long. Weight is up to 770 kg for hay and up to 500 kg for straw.

Bale formation starts at the bottom of the chamber with more rollers coming into contact as the material enters. At 40 to 50 per cent density the material starts to revolve and form the roll. The bale can be unrolled in 19.80 m. The centre is star-shaped, made up of smaller sections which is suitable for barn drying with hay. The signal for the tractor driver to start the twine wrapping is when a pressure of between 100 and 120 atmospheres is reached on the gauge and bale ejection takes 45 seconds.

Various claims are made for slight design differences in the big balers on our market. Most variations are small because there are few basic designers and sometimes only the paint appears to differ! We will come to that later. Taking the Vicon Gehlbale as an example, we have nine rubber-covered polyester belts doing the rolling, a unique spring hydraulic density control, plus adjustable, retractable bale starter which, it is claimed, ensures rapid bale commencement and a soft centre. In hay the latter point can be important, but less so with straw.

Massey Ferguson assert that in their ordinary balers the 'Suretie' knotter has 30 per cent fewer parts than most knotters. Then with their big round baler, Massey Ferguson say that there is a 45 per cent saving in twine costs. A design feature of the MF 560 is that bales are formed on steel platform rollers, as against wide belts or

56 The Tanco bale loader shown in use with the big round bale. ▶

57 The Gehl big baler at work making the round bales that have become a feature of the countryside during the harvest period. ▼

58 The International 440 pickup baler. It has a 1.52 metre pickup, (1.65 metre including gather) and will appeal where wide windrows are used and transport width is not critical.

59 The McConnel Balepacker is towed behind a conventional baler. It makes and ties 12,16 or 20 bale packages which can be loaded and stacked by a fork-lift-mounted clamp. Fed by the baler, the machine slews bales through 90°, puts them into stacks of four which are then collected in rows (5 rows of 4 bales for a 20-bale package) in the machine, and tied in a large parcel with heavy duty polypropylene twine. The package slides to the ground, pushed by the parcel behind it.

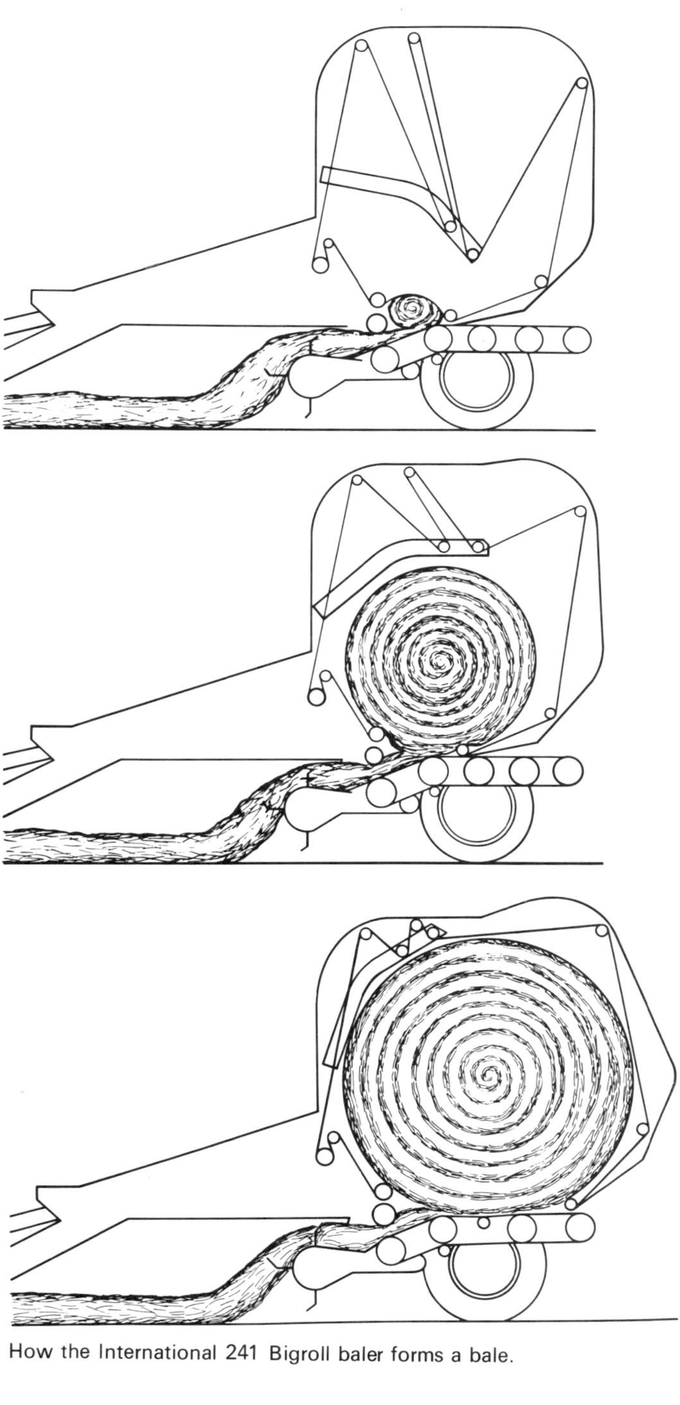

60 How the International 241 Bigroll baler forms a bale.

studded chains. The claim is that the belts are conductive to misalignment, while studded chains puncture the periphery of the bale, which may increase water penetration. Once again, you pay your money and take your choice.

Water penetration would offset one of the advantages of round bales if we used the big bale system as others do. I first saw big bales in the Northern Territory of Australia in 1972. They were left *in situ* after baling and the cattle turned into the field later in the year. It was self-feed at its best and therefore its most economical. Neither livestock farmers or livestock farming can approach the arable side for efficiency, in most cases. We all know the disadvantages, but they get a little overrated at times.

The amount of waste involved with the system that I saw must have been infinitesimal when compared with the tremendous saving in labour and machinery. I know that this does not apply to our arable lands where the stubble has to be cleared for the next cultivation; then there is the matter of our climate and our lack of acres.

I know that there is always the climatic factor of getting any UK crop, even straw, from field to safety in the least possible time. Some of the time saved can be lost at the littering or feeding end. As a result, there are now very simple machines on the market to unroll the big bales, so, provided the space is there for a tractor to manoeuvre, the farmer can add some gadgetry to front loader or rear link arms and undo the package. But that is not the end of the story. In bullock yards or sheep paddocks, it works. Amongst pens of pigs it is a very different matter.

Hence I find that most pig farmers talk in terms of retaining the small rectangular bale because that is the one they can best handle for the pig pens. For quick collection of straw for sale to straw-processing plants, obviously the big baler wins again. With any straw sale, there is just the matter of the purchaser being willing and able to accept the big bale. Thus there is unlikely to be a complete take-over by the big from the small. The two types are more likely to run alongside each other for several years to come. Then, possibly, both types of baler will disappear from our view, to be supplanted by the waferer or pelleter, or something similar, that would give the flexibility needed for using the material concerned. With baling, it is that desirable flexibility in the final usage that is missing.

THE BALE SLEDGE

The bale sledge came as a manned tool to assist the collection of the rectangular bales. This was all right until farmworkers realised

61 The Farmhand bale form used in conjunction with the 8-bale accumulator, which is one of the most widely used systems of bale handling.

62 The Farmhand Stackhand is capable of transporting up to 400 bales per hour from field to barn on a round trip of half a mile.

that the sledge was no place to be on a hot and sunny day. So the man went and most sledges are now automatic collectors, being tripped by the tractor drive to disgorge their haul in appropriate places. The principle of operation is no different to the old horse-rakes, where the man tripped each rake full alongside the last one, to make rows across the field that the carters could pick up.

Often, rather than worry about rows of bales, (especially with small fields) it is common to drag the bales to the nearest picking-up point. Always, of course, the limitation has to be the capacity of the sledge. After dumping bales in random fashion, sometimes they are stacked properly on the spot, to protect them against the weather. The tendency now, however, is to get them off the field post-haste.

Any outsider could be forgiven for thinking that farmers are mad to pick up a material twice. They could be right! But it is the same old story of getting on with the job. Of course, bales can be loaded direct from chamber to some sort of implement on wheels or skids, but it is not so prevalent as would seem logical, because of this ever-present fear of not getting a job done before the next rainfall.

There is always, as with the combine-attached straw press, the fear of trying to do two jobs at once. I would have thought it is a problem as broad as it is long. What amount of baling is done is safely garnered, whereby if the whole field is baled but none carted home, the only consolation is that it may well be fit to stack before the swath is dried out.

Bales can be catapulted by a swinging arm to the towed trailer, or they can be thrown via a pair of high-speed rubber belts. More simple than that is the long chute so that the bales get pushed upwards and backwards to what is usually a low-loader type of trailer. Like the tanker beet harvester, there is more flexibility in this system than many realise. If there is no urgency, getting the bales to some central point can be a one-man job. Either the whole combination outfit can go the discharge point, or the trailer can be unhitched and taken separately. Naturally, a four-wheel trailer is easier to un-hitch but as most trailers are two-wheel weight-transfer type, a drawbar jack may have to be used to take the weight temporarily. Some farmers use a bogie to make things easier.

RANDOM LOADING

There is normally a man on the trailer for proper stacking, but with some machines random loading could be practised. With random loading, the normal procedure is to tip the load, then fork on to an elevator. While guide arms are supposed to straighten out the odd bales that have fallen horizontal to the baler's direction of travel, it is not unusual to see someone (perhaps the boss man with nothing better to do!) armed with a fork ahead of the pick-up elevator. Remember, the baler does not even have the encumbrance of a sledge with this technique. Provided the labour is available for the picking-up team, the baler can work to its absolute maximum. As we move to more sophisticated hydraulic mechanism to pick up any bale at any angle, it will be seen that this is not a system to be lightly dismissed.

One of the simplest, cheapest and yet most effective bale-handling methods must be by means of the Lister 'Take-Put'. With one pick-up cage mounted on a front loader and another on the three-point linkage, the stacks, following sledging and hand stacking, are picked up and taken to barn, farmstead, rick or road truck. It is cheap and it works.

Rather more sophisticated are the tools with hydraulic-ram-operated gripping mechanism, but the bale accumulators are the ones to have caught the imagination in recent years. These amount to sledges with a new found adroitness. Still towed being the baler itself, bales are arranged in groups of various numbers, but a flat-eight system is common. Hydraulically-operated positioning gear may be used to put the bales into exact position. For instance, Farmhand, who have been keen operators in this field, use a front bale fork on the loader with hydraulically-controlled curves spikes that come down into the bales. The groups left by the accumulator are thus picked up and put onto stack or trailer. In fact, a man is quite capable of loading a trailer single-handed, besides using the mechanical fork to unload.

Another technique comes in the form of the McConnel Bale-Packer which is attached to the baler and ties twenty conventional bales into a pack, for mechanical picking up, similar to big bale picking. In this way the farmer gets large-scale and economical picking up, but small packages for hand handling at the livestock end.

ENDLESS CHAIN PICK-UP CONVEYOR

Firms such as Sperry New Holland market, big-time, contractor-type bale-handling systems, while the endless chain pick-up conveyor type occasionally seen on the Continent is now on our market. The latter is somewhat complicated, and after the picking up,

63 A Vicon straw spreader at work facilitating the subsequent ploughing in of the straw.

64 The Sperry New Holland automatic bale waggon epitomises large-scale equipment that is often in the hands of contractors.

the bales move in an ever-ascending spiral, giving four layers. Unloading is from the top, so there is the advantage of an unloading height of three metres, which would obviate the need for an elevator as would be essential with random loads tipped.

The Sperry New Holland bale waggon is a one-man outfit designed to load the bales from the field, transport them and stack them at their final resting place. The same firm's automatic bale waggon is what it says it is: it completely eliminates manual handling. There are now so many bale-handling tools available that it is impossible to go into the details of picking up, conveying and elevating to barn or rick.

One out-of-the-ordinary aid to conservation that I first saw at the Paris Machinery Show some years ago is the Hesston, which for want of a better description looks rather like a horsebox. This picks up from the window and compresses the material into a stack weighing 1–1½ tonnes. In the 'horsebox' the material is distributed by an air stream from the rotary paddle-type pick up and a deflector in the hydraulically-powered press roof. When the stack is ready for release from the compression chamber, the tailgate is lifted and pto-driven floor chains move the stack out of the chamber. A handler unit is available for moving the stack to where it is wanted.

TRAILERS DESIGNED FOR BALES

Going back from the sophisticated and expensive, big-scale tools of our trade, firms like Weeks, Warwick, Salop and Nelson manufacture and market trailers that are designed almost exclusively for bales. Nothing is truly exclusive in the transport zone, but the Nelson Farm Equipment Company, to take an illustration, make one trailer for the flat-eight system and another that is purposely made for one man to load and transport 100 bales unaided. The Warwick swing-fork trailer, by use of ingenious buckrake-type lifting devices, tips the collected load of nearly 100 bales from the horizontal to the vertical. There is the versatility of making reasonably weatherproof stacks in the field for later collection by the same implement, or taking them straight to their permanent storage place.

Just as there are variations in the big round balers involving belts, metallic rollers and combinations of conveyors and slat chains, so there will, as with the conventional bales, soon be many alternatives for handling at the usage end. The cheap and easy way, if one is content to haul one big bale at a time, is something three-point-linkage-mounted that grips or spears the bale. One I liked when I saw it demonstrated was the Ritchie 'Transfer Weight' transporter.

65 This forager from CTM Engineering is the in-line version loading pea haul from the swath.

66 Sperry New Holland Model 376 Hayliner baler in action. This baler features a 1.57 metre wide pickup, 'Channel Flow' feed system and high speed plunger, running at 93 strokes per minute on 540 rpm pto speed.

Perhaps it just appeals to my sense of admiration for the original Ferguson principles. This tool picks up two big bales, and using the tractor hydraulics, enables a small tractor to deal with what might otherwise be a big tractor job. Models are made for both square and round bales. Again, there is the turn from the horizontal to the vertical, thus transferring the weight on the drawbar.

As an example of the ultra-simple, as with the Take-Put—though there are many similar on the UK market—I select the Lister Big Bale Spear. You put one Spear on the front loader and another on the rear linkage and that gives you a cheap, simple and effective method of one man taking two big bales at a time to wherever they are required.

As I hinted earlier, there are a few basic designs in the round bale domain and many firms manufacturing or marketing under licence. This is very much the tendency in the big business world of today. Get a good basic design, patent it and make it in many colours. Thus, for example, we get the Vicon 'Gehlbale', which tells us that this is a Gehl fundamentally; also the Farmhand-Vermeer, which tells the potential purchaser that this is basically the Vermeer.

We have come a long way since the traumatic experiences of using the early balers in this country. Long overdue is this move to the sophisticated and the elite in the straw harvest.

PROCESSING STRAW WITH CAUSTIC SODA

Denmark claim sixty years of thinking and working on the idea of processing straw with caustic soda. At the 1976 London Smithfield both JF and Taarup brought their wares for the straw field. Now we have a real problem. The friendly invasion of new machines and new techniques gives the farmer the headache decision of deciding whether he should cart his straw to the processor or let the machine come to the farm—perhaps even to the field.

At the moment, the paramount problem for a farmer contemplating the utilisation of straw for the first time is whether the future trend will be to the giant plants mentioned earlier or to the Danish machines, or similar. Both types of machine are for chopping up the bales, metering a caustic soda solution into contact with the straw, mixing, then discharging pneumatically to where the fodder is required. Pressing the material between hydraulically-powered rollers before discharge is also practised on a larger machine. One assumes that the ultimate is cubing or pelleting, though there are the disadvantages of cost when livestock farmers are concerned with getting their feedingstuffs at a price that will enable them to remain in business.

PROCESSING MACHINES

While it is reasonable to assume that cubes, cobs or pellets will inevitably take over from a loose material, as it has done in the sphere of pig and poultry foods, the overriding factor of these processing machines being designed and much coveted for cost-cutting will weigh heavily with the mixed arable and livestock farmers. What is important to arable farmers who are virtually saddled with a material that looks hopelessly uneconomic, however handled, is that these machines have reached our shores. The technical details will change over the coming years; outputs will exceed the two to three tonnes per hour being talked of at the moment, but for many farmers it could be a profitable way out of the straw dilemma.

As in most cases, we can suppose that the two systems of taking the straw to the machinery and the alternative of bringing the machine to the farm will run alongside each other, at least temporarily. Owing to the capital investment involved it would appear that the one method may never give way completely to the other.

In 1976, on the Dutch polders, I saw an American machine at work that was almost the last word in bale handling, regardless of how the straw was processed, fed or otherwise utilised afterwards. The Freeman automatic bale waggon, as it is simply named, is self propelled. It will pick up the bales, load them, transport, re-load, or stack.

This machine is purpose-built and picks up bales as they are left by the baler regardless of angle. It has 110 bhp engine, four-speed transmission and two axle speeds. It picks up 70 bales, the stacking cage automatically raises after each layer is complete, the operator can select any tier for a cross-tie by actuating one of three switches located on a console and, at unloading, the stack can be split if necessary.

Normally the one-man-operated machine does the whole job, but the off-loading in cases of long transport distances is to road-truck, rather than to stack the bales or put them in a barn. The machine is versatile in that, with stacking mechanism removed, it becomes a fork lift. Conversion takes about two hours. Like the New Idea Uni system the Freeman has a competent look about it and makes claim to being able to load over 400 bales on to a truck in half an hour. If this big-time bale handling is going to catch on in the UK, I cannot help referring to the Uni system unit covered in Chapter 1.

Larger power units have been developed recently by Avco New Idea of Ohio for the Uni-system tractor, so that there are now units giving as much as 175 bhp and one of 138 bhp. Having seen the Freeman, which I have taken as an exemplification of that type of machine, it does appear feasible to link Uni-System with the idea. Already, the New Idea can be fitted with combine, forager, maize picker and maize sheller, so it is a logical thought to think in terms of both a self-propelled baler and a self-propelled automatic bale handler, always provided the efforts that are now being made to establish the unit over here bear some fruit.

FUTURE LOOKS BRIGHT FOR BALERS

In general terms, the future looks bright on the baler front for the reason that, with luck and suitable technology, allied to farm income available for investment, the market in the new big balers and the ton-up sort of machines for straw handling could find a market over and above what exists. With probably 100,000 balers on UK farms, and the conventional baler market recovering from the lull of some years back, it would be that the big-time bale equipment will find a niche over and above present estimates.

In essence, it means a switch from burning to conservation, thus tapping what is virtually a new market. It is for this reason that I allude to the straw harvest in this chapter, rather than confine it to balers. The arable farmers in what few wide and open spaces are left in Britain may be able to contemplate a secondary straw market after disposal or storage of the grain. As with growing sugarbeet, where some farmers give the impression that they only tolerate the root to get the tops for the dairy herd, no doubt many will want to throw the same sort of doubting into the grain and straw fray.

I hope that this is not the case, because this attitude is often overdone in the sugarbeet precinct. Grain will predominate, though we should not overlook the whole-crop harvesting idea mentioned earlier. The straw harvest could be looked on as a useful bonus. If we see the sort of bearish markets in grain that we have seen in the past—and for every bullish market there is usually a bearish one somewhere along the line—then the straw would also help. Having said that straw is usually a loser for all who dabble in it financially. I sincerely hope that we are at the dawn of a new era.

EFFECT OF CLIMATE

There can be no doubt about the need to look at bale density, regardless of whether it is the conventional or the big bales. A sobering factor, nevertheless, has to be our climate and the resultant moisture content of the straw (or hay). Very high densities from the field, as against re-baling would be up against the same problems as the waferer. There must never be any complacency when considering our weather. Funnily enough, complacency does often creep in. Many farmers will base their harvest and straw-gathering capabilities on what appertained a season before. The sensible ones will accept the best and cater for the worst!

The international farm machinery manufacturers are shrewd and logical. That is why they survived when motor-car makers were floundering. Quite apart from the fact that money invested in farm machinery manufacture will show a return comparable with most other investments, machinery makers usually choose the correct band-waggon to jump on. While we entertain an old adage about the early bird catching the worm, it should not be forgotten that many of the pioneers have received a raw deal. They may lack finance, business expertise, or just be unlucky. They can so easily be ahead of their time. This often happens.

RETAIL OUTLETS AN IMPORTANT FACTOR

It must always be a factor that the retail outlets mentioned in Chapter 1 can be paramount in the final analysis. As I have also pointed out, the machines sold are the ones that are seen from the window of the motor car or train, not those that are the figment of the salesman's imagination (and I do not want to be unkind to the sellers). Taking that as a criterion, the Norfolk-based Farmhand firm, privileged perhaps to be in the heart of the arable arena, has made a tremendous impression on this new found market. Howards (because of their Claas franchise connection) are now trying for the best of both worlds, with the square and the round.

The course taken by Claas is more logical and the new round baler is one that is different, even if the bales do finally emerge in the same shape and form as the competitors. If there is no change in mid-stream (as it were, in trends), then the demise of the big square bales could be foreseen. What a pity! They were easier to stack.

To be fair, all bales are difficult to transport and stack by old-fashioned methods. Stacks of any sort of bales are usually best in long, rectangular stacks where the air can get in and the water can run out. As in most cases, with fast-increasing labour costs, this is an area where the latest methods have to be looked at and the old, hand skills abandoned.

In the field of conventional balers I still incline towards the Sperry New Holland, followed by the International, John Deere and

Massey Ferguson. The big balers are too new for one to be dogmatic, but when I have seen them demonstrated I have been impressed with the Vermeer type used by Farmhand and Massey Ferguson, plus the neatness of the Vicon-Gehl bales. Farmhand, incidentally, have made modifications and improvements during the time of compiling this book. Other well-known names in farm machinery have come into the big bale sphere, such as Bamford of Uttoxeter, John Deere and Hesston. Again we have the licence arrangement where the Bamford and the Jones have a Welgar connection. For the big, rectangular type, there is only the Howard. The new Claas could be one for the future when we have seen more of it in the field. Its steel drums will be competing with the principles of rubber belts (which appeal to many people), chains and slats, or fingers, besides steel rollers in the base and any other variation of a theme.

Chapter 7

HARVESTING ROOTS

It is not difficult to find people who think that we may have overrated the genius of Sir Walter Raleigh and Napoleon. The former is reputed to have introduced tobacco and potatoes to Europe, while the latter decreed that the French farmers would grow sugarbeet, long after the Romans had found this maritime plant sweet to the taste.

Two out of these three introductions succeeded to all intents and purposes. The third, tobacco, was successful but, apart from isolated cases, it is not regarded as a crop in the UK. Potatoes, starchy and fattening, along with sugar, one of the most popular of produces, may have hastened the demise of many humans in a society prone to over-indulgence, but the modern world would be hard put to find satisfactory successors.

Napoleon got down to state aid for the crop of sugarbeet in 1811, while we found it necessary by 1925 to give assistance after the faltering start here in 1912. Following the first processing plant at Cantley, Norfolk, we built factories at an average rate of one a year for the next seventeen years. This was no mean achievement. More than half a century later, many of us wonder whether we should really have gone to all that trouble. To be fair, let it be said that cane sugar production, held up to be much more efficient and natural, is not without some history of official succour.

DROP BEET PRODUCTION IN UK?

Now that we are in the Common Market one cannot help wondering whether it would not have been best to have left the growing of sugarbeet to our Continental brethren, who are much better producers of sugar than we are. If the ultimate in EEC agricultural production is that each member country produces that which is indigenous to it, then possibly Britain will be told to get the countryside down to grass 100 per cent, full stop!

In the meantime, we plod along with comparatively small-scale production in units that are hardly encouraging to designers of machinery. Credit the British Sugar Corporation with bending themselves backwards over democratic principles and making justice seen to be done. But pandering to old-fashioned one-horse-size farms with one-acre beet contracts was hardly the way to get us from the horse age to the machine age in the beet field.

With the Government holding about one-third of the BSC shares, presumably we would expect caution at all times. The pedestrian struggle to an average contracted acreage of thirty-three acres, or thirteen hectares, must have thwarted a few harvester designers, For this reason, most inventions in relation to root crops seem to have come from farmers themselves. The market for, say, tractors, combines or balers has long since established itself enough to warrant specialised designers and teams of designers.

Not so with roots. These have been grown in such miserably small units that machines had to evolve from farm-orientated ideas. Until recently most sugarbeet harvesters looked as though the local blacksmith was lurking somewhere within the structure. Understandably, these machines have looked crude, cheap and unfinished. In fact, such machinery had that ephemeral sort of appearance that seemed to suggest that we may not have quite made up our minds that this is a crop for UK conditions. Nor have we.

A GOOD BREAK CROP

We produce one-third of our sugar requirements all right, but if any crop and any sector of machinery stands convicted of cruelty to soil then it is beet and the machines allied to the crop. In the days of hand-harvesting beets, it was normal to expect one's best barley crop from the beet land. Also, it was common to reduce the fertiliser application. Not only had the previous beet crop been very much over-fertilised, but possibly the residue had been ploughed in.

Even if the tops are carted or fed off, the sugarbeet is a good break crop and good for the land. Soil structure damage is no fault of the beets or the soil, it is our climate we have to blame. We come in second from bottom amongst the EEC sugar producers, but this is not too much of a disaster. Unfortunately, we do not know under mechanised harvesting conditions how much of next year's cereal crop has been forfeited.

We are inclined to over-play the value of the residual crops of tops, pulp and molasses. We boast of sugarbeet being one of our most productive crops in value per hectare yielding around 1·8 tonnes of sugar, 0·8 tonne of pulp, 0·66 tonne of molasses and some 12 tonnes of tops, according to the season. Beets are good for tap roots that go looking for moisture, especially in a dry summer. Naturally a wet season is going to bring several tonnes per hectare more tops and it will bring more beet on the lighter soils, though heavier soils can often produce a suprisingly good crop, perhaps with a high sugar content too in a fairly dry season.

AVOID BEET ON HEAVY LAND

Despite the security that heavy land gives in assuring some sort of crop in the dry years when the light lands may have virtually nothing, I have often thought that if Britain is to keep a permanent sugarbeet industry, there should be some persuasion to keep the beet off the heaviest soils. Tradition dies hard in farming. Just as many farmers accept the drudgery of cows because the cow saved them from insolvency in the 1930s, so many farmers keep beet in the rotation for no other reason.

This may apply to potatoes to a lesser degree, but small areas of the humble spud are often grown for the main reason of producing things that fit the dual purpose of sale or usage. Few, if any, households get by without the potato. Potatoes are not just a human food; they are pretty useful pig food too.

It is a laborious job to make one's own sugar from the beets, so I doubt whether any farmer has ever grown beet with that thought in mind. What has affected the agricultural industry in general and machinery designers in particular is the over-kind demeanour that has tolerated foolishly small units of production.

Nor has the man on recalcitrant clay who persists in growing sugarbeet helped. The crude, country blacksmith look persisted in beet harvesters years after the combine harvester looks sophisticated, refined and elite. It was patently obvious years before it happened that beet harvesters had to follow the road already trod by the combine harvester. Sadly, ridiculously adverse land conditions, plus a spattering of ridiculing farmers, could, have killed off a few embryo ideas.

FARMERS' VIEWS AND DEMONSTRATIONS

I well remember the argiculturalist of a sugarbeet factory telling me that he thought it wrong to persist with demonstrating in adverse conditions on heavy land because good prototypes had been killed off that way. I believed him explicitly. Farmers will gleefully condemn a machine at both ends of the scale. If the BSC hold their demonstrations on light land, a machine working well may still be condemned because the onlookers assume it to be useless on heavy land. Often they are right. Just as often they are wrong, unfair and presumptious.

If a demonstration takes place on heavy land and gets rained off, or conditions are unfavourable for the machines, then the machines get maligned just the same. In fact, one of the great credits due to the BSC is the excellence of their demonstration arrangements. The policy of financial deductions for development and research from the sugarbeet cheques, along with the national demonstrations showing machine developments in working conditions, is something that would have stood any crop section of farming in good stead.

It could be that as the years go by and machinery develops, there will be no need for working demonstrations. Even then, the crop advancement and the seeding and spraying techniques may well still be on show, just as there are cereal plots to be seen at the invitation of the Ministry of Agriculture, seed purveyors and fertiliser manufacturers. I doubt whether any official or quasi-official body (such as the BSC) has done more to educate their growers by practical demonstration. At the same time, it could be nothing more than a help to the people making the machines for the job.

Nevertheless, a much larger average contract unit, more row width standardisation and no beet on the heaviest clays would have done the machine manufacturers just as much good. The first complete beet harvester that I saw in this country was in 1939. However, it was not until about 1950 that even medium-scale production of harvesters was into its stride and farmers sold on the idea. It was not far short of 1970 before we could say that machine harvesting was 100 per cent efficient.

EARLY MACHINERY DEVELOPMENTS

The in-between period was brief, after the horse lifter, we flirted with the Ford-two-row spinner, which, let it be said, was not a failure. It was overtaken. We played around with imported squeezers and toppers and even the great Harry Ferguson wiled away some of his time on sugarbeet equipment. A down-the-row thinner for the spring and a single-row tipper for the autumn were tried but with precious little success.

Using a Ford spinner, putting four rows into one, then topping by hand, had a brief existence. Soon the Catchpole Cadet, the John Salmon and the Peter Standen were to dominate the scene completely. These three, especially the Cadet, converted farmers to mechanical harvesting, transformed sugarbeet growing and, while the labour was available for the cartage involved with side-delivery machines, became a success story in the 1950s and 1960s.

Now, following the inevitable tankers and the equally inevitable self-propelled models, we rather appear to have gone full cycle. Plenty of Catchpole Cadets still lurk in muddy stackyards around the farm awaiting the call that comes with each bad autumn. When the tankers and the big, new giants get bogged down, then out come the old stagers for side-delivery work. Yet some of the modern harvesting techniques involve the return to side delivery and a large team of men.

There have been times during the years of development when the livestock farmer, who insists on a machine that leaves the tops suitable for collecting for feed, appeared as something of a drawback to the designers. At times, the real arable farmers have had to wonder whether this crop was being grown for the 16 per cent of sugar that we were likely to extract from it or for the by-products. As we went on to tankers, then self-propelled tankers, demonstrators were almost ostracised and designers blasphemed if the tops were not in perfect condition.

CONTINENTAL HARVESTERS

Then came the Continental five-row and six-row, three-piece harvesters. Stones and clods could be a little mixed up with the beets and the tops, but it proved a point that designers were out for economical harvesting, not preserving the by-products.

Neither were the Continentals too fussy about the top tare. After all, they had not been brought up on a recipe of neatly topped beets. Flail topping had been the order of the day, with 15 per cent more of the beets going into the factory. The livestock man could have what was left, if anything! Alas, the unbending heroics of our beet hierarchy would not stand for that. So the friendly invaders from across the English Channel had to add knife toppers at the rear, with the original flail toppers up front.

MACHINES TO USE

Now if one looks into the economics of beet growing, they will find that anyone with about 28 hectares would do best to own the three-piece machine. Even at around 22 hectares it pays to have the three-row, two-piece machine, such as the Standen Multibeet. Up to 20 hectares it would appear that the popular, one-row tanker, whether trailed or self-propelled, is the best bet. In all these figures there is a fair amount of overlapping.

There are plenty of farmers growing over 28 hectares of sugarbeet who have stuck to the single-row tanker. Not without good reason. Continental farmers go for one machine topping, the next lifting and the third gathering the beets, doing come cleaning and delivering into trailers side-delivery. Just like the little Catchpole Cadet twenty years earlier! Peter Standen accomplished the great British inbetween act with the Multibeet. This outfit has one machine doing the topping, with another lifting and delivering to the trailer alongside.

I suppose that one of the difficulties in making the final machine a tanker in either system is that with multi-row harvesting the tank would get full so quickly. The almost continuous job of emptying could make the job little different. Multi-row harvesters are now tending to get into co-operatives, with neighbouring farmers combining for the most economical harvesting.

The problem of delivering a few stones to the factory along with the beets has not been completely solved. The farmers who grow beet with one and a half eyes on the tops and just half an eye on the roots have often got over the problem of having perfect cattle feed by foraging off the tops ahead of the harvesters. Rising costs in the feedingstuffs market have made beet tops a sound proposition and somewhat vindicated those who insisted on harvester design that catered for top saving.

WATCH PRIORITIES

We must still be careful to get our priorities right. At the 1976 BSC autumn demonstration it was reported that 26 per cent of the tops get utilised for feed. We assume that the figure will go up only because of the rising price of feedingstuffs. The world sugar market was not always pleasing to producers in 1976, though there is never any guarantee that a market trend goes on for any specified time. Sugar could go up in price. Feedingstuffs could come down. Anyone who has stocks and shares or gilts and equities, who has to anticipate bull markets and bear markets, knows that nothing in this world is perfect. That includes beet machinery.

For the above reasons and because few machines have become established in that particular region, I want to dispense quickly with the topping, foraging and picking-up machinery. If the UK farmer wanted the pick-up waggon, he could have had it years ago and long before someone got around to realising that it would pick up beet tops. Purely as examples of what is used I mention the Agrostroj, the BTS-Pottinger, the Krone, the Mann-Claas, the Melotte and the Strautmanns. There is also the CTM forager-type top-saver from King's Lynn, Norfolk.

67 The Armer-Salmon beet harvester with its feature of topping after lifting is the only one of its kind marketed here.

68 The Standen Rapide Tanker, which can also be fitted with top saver, is considered one of our most successful harvesters.

I do not want to ridicule or condemn, but some machines which cost good money may exist because some organisation says they should. Beet cleaner-loaders are perhaps in that category. To be fair, dirt has become a problem for the factories. There are not enough gardeners or house builders in Britain to utilise all the soil that the factories have to dispose of. Farmers will take the lime sludge as a near-gift, but that is about all.

Hence the big push towards the cleaner-loader and the concrete base. The cement-laying exercise is correct and well worthwhile. In fact, on heavy land, it is now virtually a necessity. As the same base can be used for farmyard manure, lime, sludge lime, drainage stone and so on, loaders are difficult to justify on pure economic grounds, but they are here and almost certainly going to stay.

It is a good, logical sort of argument. Keep the valuable top soil on the farm. Reduce haulage costs. I can remember the time when some landlords would not let their tenants grow sugarbeet because, they said, some of their assets would be carted away, never to return. Thay may have been good feudal folklore, but most landlords would now take the opposite view. They would rather see sugarbeet in the rotation than have tenants growing all-white straw crops.

We all know that if some dirt is not taken to the factory, the haulage would be that much less. As many farmers run their own lorries, even that may be overrated. The salvaging of top soil is a good theory. Whenever I have spread this soil back on the land, I have found it a certain way of making sure that the soil grows little or nothing for the next year or two! It is like having a cesspool emptied, with modern detergents much in evidence. It is very much the same with sugarbeet dirt.

So farmers are encouraged to clean their beets at farm level because an over-supply of dirt at the local factory is an embarrassment. Farmers, being co-operative and conscientious, rightly involve themselves financially by investing in those rather over-priced machines. Beet cleaners are simple and uncomplicated. Any good blacksmith should be able to make one, but they do not bother. Nor do the large firms. It is a job for the in-betweens. The good thing about these machines is that the beets have got to be loaded anyway, so they may as well be cleaned at the same time.

FORAGE TOPPERS

As with the newer forage toppers, I do not intend to dwell on the subject. Some firms have long since given up making them. The few left in this market can cope with it satisfactorily. Incidentally, the move to much larger capacity machines has not been because farmers are necessarily buying them. There has been a tendency to put out the cleaning, loading and carting job to contractors at an all-in price (except, that is, where farmers run their own trucks). Contractors cannot have their men standing around farms too long, so they go for a large cleaner-loader which gets towed round to their clients, as and when wanted. Often the tractor pulling it and also loading the beets is the ordinary farm tractor, plus fore-loader, but it is sometimes a more sophisticated fork-lift truck or industrial loader.

We accept hand pickers in the potato field, but loaders that come onto the market with provision for hand pickers to assist the cleaning either have to do the job minus the human hands or they are short-lived.

There are many farmers, especially on light land, who still survive the trials and tribulations of beet growing with a front-end loader loading straight into the lorry. If I were farming light land, I would do the same. It is little use reflecting on the stone problem, because the only stones that will be extracted from the sample are the small ones that will go through the trace or the cage, as the case may be—assuming, of course, there are no hand pickers on a cleaner-loader. How a farmer can justify them, on top of his necessary costs to grow sugarbeet, I do not know. I cannot remember seeing it done other than at demonstrations.

Until recently there were just two types of cleaner-loader—the revolving cage and the scrubber chain. I have had good, bad and indifferent results from both, but I prefer the cage type. The scrubber chain only seems to be doing what could have been done on the harvester, except that the dirt may have dried out in the intervening time and then comes off easier. The Armer-Salmon Universal has put a better face on things so far as scrubbers go, but some of the earlier models left a lot to be desired. The Todd, from Root Harvesters of Peterborough, has always been a useful machine at a realistic price.

CTM of Kings Lynn actually offer three models of varying capacities. Claimed outputs on cleaner-loaders is something that I take sceptically. The reasoning is this. If the maker says that the machine will load twenty tonnes per hour, that may be so. It will depend on how quick the operator on the loading tractor can work and how large his bucket or root fork is. More important, if the beets go up into the truck too quickly, it may be that the cleaning mechanism will not have time to take any dirt off. In muddy conditions on clay land, I have seen beets rolling the clinging mud onto themselves as they go through the machine.

The cleaner-loader that has stood the test of time better than most is the Cooper Tareaway, from Cooper's of Driffield.

Among the new ideas in this sphere is the use of more up-to-date mechanism via hydraulic power, while the ridiculous little petrol engines that were used for years after they had outstayed their welcome have given way to diesels. The pto-driven machine did not last either. Considering the price of a small diesel engine, to have a tractor standing around wating for the lorry to come did not look a very good proposition. As happens, of course, many farmers had an old stager that became the permanent source of power for that type of loader.

REVOLVING CAGE ON FRONT LOADER

Another new idea is the revolving cage on the front loader so that the dirt is shaken out as the tractor or forklift truck travels to the truck. That way the dirt gets spread out over a large area, as against some other types that deposited it in one place, that is, underneath the machine. Of course, many farmers who farm near the factories are able to send their beets from harvester to trailer, then straight into the factory. On the Continent this method of going straight in with farm trailers is much more prevalent than here.

While the ordinary farm loaders suffice for most beet-loading operations, there has been a tendency to use rough terrain forklift trucks in some cases. Then there is the slew loader like the McConnel Power Arm with beet grab. Volvo, from Sweden, have been seen at sugarbeet demonstrations with a beet basket on their loader.

Another innovation from Sweden is the 'Quickie' drive-in front loader from Lawrence Edwards.

Back at the harvester end, which is the important part of the job, we find that the British are struggling. Despite talk of our advantage in average farm size, it has to be remembered that on the Continent there is a lot on contracting in the beet field. So the big-time machines are not so surprising as may at first appear. One of the best that I have seen working is the Moreau 5-row complete harvester imported by Fords Ltd., of Rackheath, Norwich. Topping, lifting and delivering five rows to trailers running alongside, I have seen the Moreau polish off six hectares of heavy-land beet in a day. Commercially, I mean, not at a demonstration.

The same firm has a tractor-mounted 5-row defoliator for front fitting, with the 5-row lifter at the rear end of the tractor. I find that many farmers are turning away from the three-piece machines and looking hard at the two-piece systems. I do not blame them. Especially, it seems, the cow-keeping farmers who want their tops are

69 The Standen Multibeet two-piece harvesting system in action, with the trailed topping unit in the foreground.

70 Standen Multibeet flail topper.

71 The Standen Solobeet Cyclone which can be fitted with a variety of tractor makes. The power unit is removable and can be re-assembled for ordinary work.

72 The Ransomes Powerbeet built round a Ford tractor.

73 Artist's drawing of Ransomes 2-row self-propelled beet harvester, which is powered by a 74 hp engine and weighs over 6 tonnes.

74 The CTM Super 120 sugarbeet cleaner/loader with throughput up to 2 tonnes per minute.

75 The Herriau 3-part harvesting system from France, imported here by Stanhay, seen loading beets direct to lorry, with the topper in the background. Between these two machines comes the lifter-windrower.

76 The Whitsed Vulcan 2-row potato harvester at work.

beginning to like the look of machines that will put the tops where they are wanted and gather the beets in a twice-over operation.

Herriau, from France, are the people who have really got the British onto the three-piece system, so far as the large farmer is concerned. Catchpoles tried this many years ago without a lot of success. The difference now is that tractors have got big and powerful. We have hydraulics and the 1000 rpm power-take-off, besides the old 540 rpm. Not that we have to use it, but along with hydraulic engagement and such refinements we are in a better position than we were.

PURPOSE-BUILT MACHINES

One of the most useful purpose-built machines in the beet field has been the Kuiken, from Holland, a 6-row self-propelled model that is usually accompanied by the Franquet pick-up loader. Much modified, this harvester was imported into the UK in 1976. The Czechs have entered our market with a 3-row topper and a 3-row lifter known as the Agrostroj. The Amac from Benedict is a 3-row complete harvester outfit. Besides the Herriau, we have the Matrot 6-row unit in a three-machine format.

Salmons of Essex merged their harvester with the Armer from Eire and we now have what was the Salmon 21 self-propelled model, plus the well-known single-row Armer, a model with an hydraulic top-saver as an extra and one that gets called the Three-in-One. This is another built-around-the-tractor type; it is fitted with a stone extractor and is self-steering. In fact, some sort of self-alignment to keep harvesters on the row properly is becoming a feature. It can be a simple two-pronged welded assembly, as available on some of the Standens, or something more scientific.

One of the modern economic problems with self-propelled harvesters is the cost of the power unit. We have the example of the Ransomes Hunter, a 2-row machine that looks engineered for the job.

Standens, of Ely, for years kept prices down with the Solobeet that utilised a tractor on the farm, or a secondhand one bought in. Either the tractor could be left attached, or with some amount of work it could be detached and re-assembled in its original state. What it really amounted to—and the model known as the Solobeet Cyclone is still popular—is that the farmer who fitted a new tractor had to detach it to get full value, but the man who fitted a half-price unit could afford to leave it *in situ*.

THE TOPPER UNIT

Since we virtually discarded the hand hoer in the beet field, the uneven braird that we get from planting to a stand with monogerm seed has focused some attention on the topper unit. Toppers have never been very efficient; to me, they have always looked heavy and clumsy. On the other hand, if we had a light unit there could be bouncing that would cause the under-topping that could be as unpopular as the over-topping a unit weighed down with mud may do. I know that there are all sorts of compensatory spring arrangements and the man with his 'spud' to spade the dirt away, but my experience is that topping efficiency is much upset by clinging mud.

So the NIAE and now Standens have looked at what they call the lightweight topper. When I saw the units in work, their performance was mixed, but I thought them better than some that have gone before. The open-centre, lightweight design can hardly be wrong for most English conditions.

It has to be remembered that sugarbeet growers get penalised at the factory for over-topping and under-topping, in much the same way as they get increments and deductions in price per tonne for sugar contents over and under 16 per cent. When extreme over-topping takes place, the farmer is robbing himself and sending less root weight to the processing factory. If the tops are utilised for a herd of cows, there is always the factor that the loss would be less. Judging by the amount of bad topping that is to be seen during any season, one even wonders whether the livestock farmer turns a blind eye to the topping. If the tops are ploughed in, then the over-topped portion becomes expensive manure!

Once in the field, there are three jobs to do in a difficult environment. At least, for the most part, in the UK beet-harvesting conditions are going to be laborious. First, there is the bulk discharge and with the side-delivery types, single row or multirow, we have the continual discharge. Of these operations, either harvester operators are less proficient at setting the topping mechanism, or this is the weak link in beet harvesting. Once the balance has become upset, even through the necessary expedient of fitting a sharpened knife, it can take hours to get everything just right again.

BEET HARVESTER KNIVES

I do not want to labour the point of farmworkers being bad knife sharpeners for the most part, but this judgment applies to beet harvester knives more than in any other department. I tested some knives for NIAE in 1974–1975 using two different heat-treated steel knives, but any advantages over what were already on the market were minimal. The reason why knife sharpness or bluntness is important is because it is this very factor of the cutting profile continually changing its degree of effectiveness that throws the topping unit out of alignment.

Where the beets meet the knife face which to all intents and purposes is in the centre, most wear till take place. The operator often compensates for the blunt knife with re-adjustment and makes a major re-alignment at the moment of affixing each newly sharpened knife. If this is not done, then over-topping takes place at the commencement of the operation, with severe under-topping just before the next sharp knife is fitted. Loss in the UK crop through bad topping or incorrect adjustments has been put as high as 4 per cent.

Total crop loss on the average farm can be anything between 2·5 tonnes and 3·7 tonnes per hectare. There is no such thing as harvesting beet, or any other crop, with a nil loss. With a beet-topping knife that will spend a lot of its life blunt in the centre and sharp at the two ends, there will be sudden and costly overtopping when a driver loses concentration, or where there is a slight demeanour in the drilling. Thus the basic problem of beet topping is, ironically enough, to start off with a blunt instrument and to stop all variation along its cutting profile and throughout its life. There is the possibility of topping done with a wire, not unlike the grocer's cheese cutter.

FLAIL TOPPERS

Flail toppers are coming more into vogue, but there has always to be the precise cutting-off of the top to suit the British factories. The small amount of additional sugar from, say, 15 per cent more bulk is uneconomic. The farmer is the loser. As we move onto more general use of automatic guiding principles on beet harvesters, topping and lifting should improve. Cleaning is something that we accomplish in direct relation to conditions at the moment of lifting the beets. The scrubber chain and the spinning cage mechanism that may also throw away a few stones for us have both served us well.

If the exigencies connected with beet harvesting were stable, it would be something in our favour. Unfortunately they are not. A frost, a shower of rain, some snow or even some fine days, change soil conditions alarmingly. Almost continually, the operator has to set up his machine as the changes take place. This loses a lot of time. Often, I am certain, operators balance one factor against the other. You lose time with the re-adjustments to do a better job and possibly invoke the wrath of the boss because insufficient hectares have been harvested. Or you press on, knowing full well that a mediocre sample is being produced but at least some beets are being got to the roadside.

Back to the vexed question of topping, I feel bound to say that if there has to be a choice between over-topping and under-topping, I would normally settle for the latter. The returns from the factory might show a huge top tare figure, but it all has to be kept in perspective. Beet are not even a good green manure crop ploughed under. Any farmer who has had to plough under a part of his crop in a disastrous year will relate the story of how the cereal crops did not come up to scratch in the following year or two.

THE LESS SOIL LIFTED THE BETTER

Fundamentally, we come back to the dictum that the less soil lifted with the beets the better. Following the shares of the horse lifter and all sorts of lifters and prongs, we settled mainly for the lifter wheels for years. Now we go for all types including vibrating the shares.

As in so many cases, our soil variations mean that what suits one farm may not suit the one down the road a mile or two. Not that this attitude should be overdone, because too many farmers think that his land is different to everybody else's. There are lots of land variations but making too much of this problem only leads to a conglomeration of variants offered by manufacturers. Any farmer, beginner or otherwise, moving to a new area has only to look over his fences. This is one of the great advantages that farmers have over industrialists. The machines a leading farmer in a certain area buys today are the ones that his neighbours are most likely to buy tomorrow.

POTATO HARVESTING

Many of the remarks made about sugarbeet harvesting apply equally to the potato field. Use of the basic spinner is diminishing. It was overtaken by the elevator type of digger, which is now being superseded by the complete harvester. In the same way, the bags are giving way to bulk methods. All this is inevitable owing to increased labour costs and the absence of casual labour to pick up behind the spinner.

The potato boom of 1975 and 1976 (in price rather than yield)

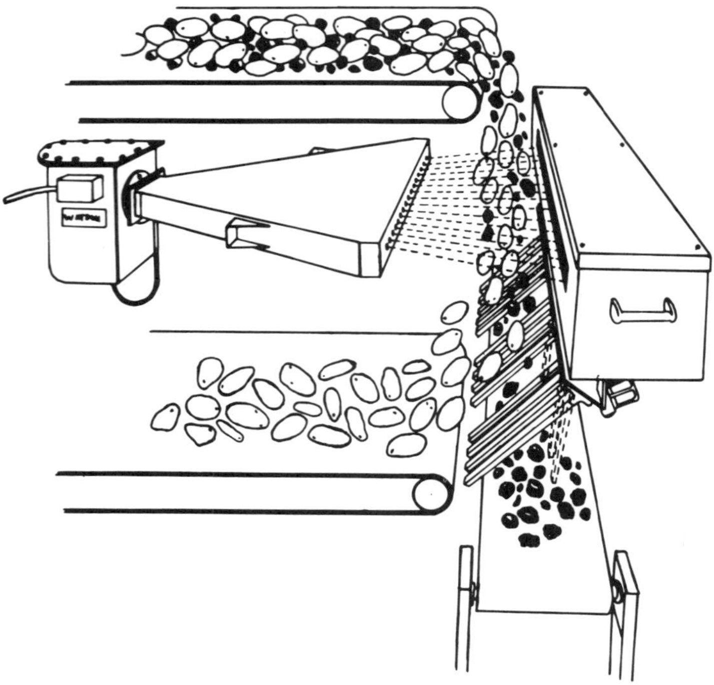

77 Diagram showing X-ray sorting system for Whitsed Automatic potato, vegetable and bulb harvester.

78 The Standen Startrite potato harvester in use with hand sorting.

79 The Grimme self-propelled potato harvester, powered by a Deutz 100 hp 6-cylinder air-cooled engine.

80 The Benedict Amac 2-row self-propelled potato and vegetable harvester.

should have left a fair margin of profit to move from elevator diggers to complete harvesters. Not that the farmer has to spend potato profits on potato machinery. As I am at pains to point out, the profit from one crop is not bound by any rules to be used for purchasing machinery for that particular crop.

I suppose that the ordinary type of spinner or digger, pto driven, is an uncomplicated piece of machinery by today's standards. As with beet, there is always the golden rule of lifting only as much soil as is necessary to get all the crop. Also, like the sugarbeet crop, easily-worked soils are a much better proposition than clay. A modern tendency has to be towards rubber and plastic coating of the metal parts that are liable to damage the potatoes. Obviously the lighter the land, the easier it is to shake the soil through the chain web, while the elevator type will display the crop better for subsequent picking than the spinner. Moreover, elevator diggers are multi-row in some cases, so that two rows can be put together for a tidier row of tubers across the field.

Over-agitation of the crop will, of course, cause severe bruising and upset the keeping quality. Just as the malsters do not want their barley skinned, nor do shops and processors require bruised potatoes. As with the old side-delivery models of beet harvesters, our unreliable climate takes farmers backwards through the machinery age in very bad autumns. The complete harvester may have to give way to the elevator digger, or the digger may have to make way for the spinner.

It stands to reason that the complete harvesters are more suitable for the main crop, because of the bulk to be dealt with. The earlies, or even second-earlies, may be picked by hand because the price is high. As complete harvesting develops, it seems certain that, especially on the small farms, no farmer will invest in a complete harvester just to harvest the main crop.

A SMALL-SCALE OPERATION

The trouble is that potato growing is often a small-scale operation. While many large-scale operators engage themselves in planned potato production in the fenlands, either for processing, wholesale or retail selling, even large farms seem to revel in a little sideshow. Potatoes fill the bill so well because they are easy for cash sales; they can also be consumed by the pigs if the trade is slow and every household can get through a few tonnes in a year, if need be.

In line with all agricultural produce there is such a thing as pricing goods out of the market. Whether beef, potatoes or anything else, this is a lesson to be inculcated by producers. Farmers incline towards the view that roast beef, bread and potatoes will go on, whatever happens. Consumers can be fickle, both on quality and price. There is never a guarantee that human nature will not change, however stable we consider ourselves to be in the UK.

During the shortages of 1975, many potatoes reached the kitchen in a state that made the onlooker contemplate the long-term effect. Bruising, cutting and any sort of damage will only be tolerated during extreme shortages.

FARM-TO-CONSUMER SELLING

Farmers will always moan loud and long about the middlemen who take what looks like a disproportionate amount from produce that has been produced at some amount of risk. Crops like potatoes, fruit and eggs can be the exceptions to the rule. There is much talk of direct selling. Notice boards point to potato stores and the price is exhibited for the passing motorist to be enticed into filling his boot. It is a good line for farmers to take during shortages. It sounds good and it works. Often it works all the time, shortage or no shortage. But I have a feeling that this attitude could go full circle.

We have all bought rubbish at what looked attractive prices compared with the retail shop. There have been plenty of arguments and complaints. Direct farm-to-consumer selling works with potatoes, but it relies on quality more so than sales through a wholesaler. There is no middleman to blame from both ends! Thus the importance of harvesters that put potatoes into store or bag without damage assumes tremendous importance both from the practical point of view and as a public relations exercise.

As in the beet field, farmers give the potato machinery designers a few problems by asking them to cope with all sorts of row widths. Some manufacturers make different models for different row widths. Pulverising flails can be of different lengths and positioned to suit row shape.

Elevator diggers are often equipped with chain web for stony and sticky conditions, but are of the belt type for abrasive soils. Twin chains or belts are sometimes fitted for very stony or hard conditions. Diggers can be trailed, mounted or semi-mounted—and castor wheels at the rear get used, too. One of the surprises to me is the lack of use of castor wheels, especially with the farm sprayer that is just over the limit for carrying on the link arms.

We are in the age of vari-speed for the varying forward movement relative to machine speed and crop conditions. Machines that use hand pickers on the machine may balance the weight distribution with the subtle placing of the pickers' platform. A rubber belt sorting table ensures both safety and ease of picking. There are often optional rod links for the variations that inevitably occur in British conditions. Using hydraulic power for adjustment final unloader mechanism is now common. Potatoes lend themselves to pallets and containerisation. Boxes on the machine or used in conjunction with hand pickers, later to be picked up by loader or fork lift truck, are now in vogue.

Harvesters are sometimes offset so that the tractor does not run in the crop, with the separating area in full view of the driver. There can be a choice of picking stones and clods from the potatoes or vice versa. Makers will obviate damage with low drops and patented nesting features. There are arrangements for hydraulic steering on the rear wheels and a ram on the drawbar to ease entering the crop at the headlands. For machines with hand pickers there is the comfort of a canopy and seats, besides shaped sides on the picking platform with toe room.

Many of the modern machines will work reasonably well on side-hills, and the haulm pulveriser may be optional for simultaneous haulm removal. An example of a machine with a fairly low initial cost and which has shown low damage at demonstrations where damage has been measured is the little Ransomes-Faun from Norway. Rightly or wrongly, there has been a tendency to drop performance testing at national demonstrations, both with potatoes and sugar-beet. This was last done with potatoes in 1970 and the last damage testing was in 1972.

It stands to reason that farmers will bellyache about such matters. It is arguable from both ends. The manufacturer can be unlucky on the day with his machine. As at farm level, it can just as easily be the operator who causes a breakdown of the machine as a shortcoming in the machine itself. Even on the demonstration site, one manufacturer may be unlucky with the plot that he was given. The farmer will argue that the machine should cope with such variations and that he still wants to see some figures.

For one thing, I doubt whether the average, deep-thinking farmer will make his final choice on the figures of one demonstration alone. Another point is that he would be foolish to do so. Because of the factors mentioned, a machine can have a good day or an unlucky day. To judge a machine on a one-day performance is not good for either the maker or the user.

Personnel connected with the demonstrations for both potato and sugarbeet machinery have inclined towards the view that it could be better to let the growers sum up for themselves, rather than rely on one-day figures. They could be right; there are good arguments, as I have indicated, from both sides of the fence. The only reliable and worthwhile testing, so far as I am concerned, would have to come from some impartial body and over a reasonable period. The only organisation that springs readily to mind is the NIAE.

We then come up against the historical evidence of how farmers failed to support the tractor-testing scheme. We are now faced with compulsory levies for research and development, as well as training. Frankly, I know of no other way to get farmers to support such things. Getting money for training was no more successful than getting money to issue test reports on tractors, as we all know. In both cases, it added nothing to the farmer's public image. It almost certainly took something away from it.

TESTING PROTOTYPES

We must not overlook the disadvantages of prototype testing of machines by an official or quasi-official body. Manufacturers want secrecy and they are entitled to it. Sending a prototype beet or potato harvester to Silsoe for a season of testing on different soils would of necessity see it taken around the countryside. Manufacturers themselves have to put the secrecy aspect at risk with their own testing, of course.

The Massey Ferguson 711 potato harvester was an example of a manufacturer co-operating and taking up a NIAE development. Secrecy during the embryonic stages is difficult, regardless of whether the machine is in the hands of a manufacturer or such a body as the NIAE, but it must inevitably be more difficult with the latter. This can only militate against the policy of sending all prototypes to an official testing station, though I feel certain that this is the very thing that farmers desire.

Another example of development and manufacturing co-operation is the X-ray and electronic potato-sorting system as used by Root Harvesters of Peterborough. With due regards to the Scottish NIAE Station for their work in the matter, I refer to the tremendous cost saving if human hands can be eliminated from the sorting table. The mixture of potatoes and trash is channelled between a sensor and a source of X-rays. Stones have a higher resistance to penetration by X-ray. The sensor operates a pneumatic mechanism which causes deflector fingers to be withdrawn, thus allowing the unwanted stones and clods to fall back to the ground.

A more simple way of removing stones is by means of a roller covered by nylon bristles that are soft enough to allow the stones to penetrate, while the softer and lighter potatoes do not. Pairs of rollers may be used as clod crushers, but obviously they can deal only with the large clods because if set too close they would crush the potatoes as well. Clod separation can be dealt with by a studded belt running slowly or a smooth belt running much faster. The hope is that the round, more mobile potatoes will tend to roll down a sloping and moving belt, while the rougher clods and stones will travel with the belt, to be picked off by hand pickers or returned to the soil.

No matter what system is used, between X-ray separation at one end and a gang of expensive hand pickers at the other, trash will always be a problem. That is to assume that most farmers grow some trash amongst their crops! With potatoes, there is the haulm anyway. Spraying with a chemical desiccant some days ahead of harvesting is one method of dealing with the problem, besides getting some blight control at the same time. A complication in the potato field is that earlies, second-earlies and main crop may require different treatments.

One does not want to stop the earlies growing a moment earlier than necessary in a country where it is often difficult to get a worthwhile yield. So the mechanical methods of haulm pulverisation are likely to stand up to the chemical challenge in the earlies and second-earlies. Most harvesters are equipped with such things as revolving slashers to clear the top of the ridge. Haulm destruction can be a separate operation with a machine such as the Wilder, to take a common illustration or, of course, the pulveriser can be on the front end of a tractor with the harvester at the rear.

SELF-PROPULSION AND OTHER DEVELOPMENTS

As in the beet field, we now see potato harvesters moving into the arena of self-propulsion. There are now multi-row harvesters and single-row harvesters, tractor-drawn or self-propelled. The typical examples are the Grimme machines from Germany that are brought in by Richard Pearson of Lincs.

Lockwood Graders (UK) Ltd., have given the farmer the oportunity to use X-ray separation at the store, in conjunction with an unmanned harvester in the field. In other words, the unmanned harvester is not generally accepted as 100 per cent infallible, but any deficiencies can be put right at the store.

The Loctronic precleaner, which incorporates X-ray separation of stones and clods from potatoes is fitted with a haulm extractor, undersize and over-size spool grading mechanism, besides soil discharge elevators. Boughton & Sons of Amersham (well known for their winches) have built a tractor on an International Harvester chassis, with a claimed all-time low-in-ground pressure of 1·8 psi.

Built to a Century Fabrications design on the I-H chassis, the tractor would likely cope with the fenlands and the boggy areas of Ireland better than most if and when it comes to a rescue operation to salvage the potato crop. I visited the peat fields managed by Bord-Na-Mona (an Eire Government Authority) in Ireland some years ago. In those days they relied on such things as half-tracks, mainly on Massey Ferguson tractors. I deliberately mention this tractor here rather than in Chapter 1 because it is a specialised tool and the finance involved tends to restrict it to such crops as the remunerative potato and the peat that fuels electric-generating stations.

The crux of the issue is the 1·8 psi ground pressure. There is no doubt, now that potatoes are big money, that the thought of leaving any part of the crop unharvested because of bogged-down machinery is a nightmare for the grower who has financed the crop.

HYDROSTATIC TRANSMISSION

Hydrostatic transmission is in the fringe areas of potato machinery. There can be little doubt the hydrostatics, four-wheel drive and exceptionally low ground pressures are items that have to concern the designers of both potato and beet machinery in the immediate future. Offset machines have many advantages, but there is always the disadvantages of crabbing, especially if power or weight is limited. Four-wheel drive and tracks have so many advantages when the going is slippery and tough. I would go so far as to say that there has been insufficient attention to traction in this sphere.

Fenland farmers and heavy-land farmers who are involved in potato or sugarbeet growing need machines that will traverse the ground after a cloud burst. So often with both crops, it is a salvage operation. In this matter of ground adhesion, we are only at the beginning so far as these two crops are concerned.

In making my preferences in this section, it has to be stated that it is not a matter of a straightforward choice between makes of machinery. This is an area where the British manufacturers look to be losing ground to their Continental competitors.

It is now very much a matter of systems rather than mechanisation. We may want to harvest one row or several rows. We may want self-propulsion or the cheapest that money can buy. It is not like it is in the combine harvester province, where there is no suggestion of anything other than a self-propelled machine. If the large and expensive Continental machines capture this market, may be the British farmer will emulate his Continental counterpart by also becoming contractor-minded. As always, with contract farming in the UK, there is the vexed question of our climate. The machine, however large, can only be in one place at a time.

British farmers cannot claim to be either good contractor material or good co-operators. No one has the right to blame them for it. It is very much a weather factor, added to the yeoman independence that many admire in our country people. There is one good point that should be mentioned here. After the Milk Marketing Board, I rate the British Sugar Corporation as one of the great successes of our farming time, with the Potato Marketing Board not too far behind.

SELECTING MACHINE FOR THE JOB

If I were a medium-sized sugarbeet grower, I would select the Standen Rapide, with or without top saver. After that, or on really heavy land, I would choose the Ransomes 33. For stony land there can be advantages with the Armer-Salmon, but do not forget the vulnerability to frosted or small tops. If I wanted a two-piece harvesting system, I would favour the Standen three-row, with flail or knife-topping according to taste and future destiny of the tops. If it had to be the three-piece system for big acreages, I would take the Herriau. For something really big in self-propulsion, but delivering to trailers, there is the Moreau AT6. For a big-time tanker, the Ransomes Hunter.

If I had to select a value-for-money buy, it would still be the Standen Rapide Tanker for its performance under most conditions.

The spinner side of the potato machinery market is dying, though the growth in complete potato harvesting machines and techniques has been rather slow. Potato harvesters do now account for over 60 per cent of the national acreage, along with elevator diggers that harvest most of the remainder. Something like 50 per cent of the market is for new machines that fall within the spectrum of partially manned and unmanned harvesters. Obviously, the breakthrough with stone windrowing opens up a new vista for purveyors of complete harvesters on lands that they have rated as difficult until now.

Harvesters favoured by many growers are the Whitsed, the Grimme and the Amac machines, while many smaller growers would rightly choose machines of the calibre of the Standen Startrite and the Ransomes Faun.

Chapter 8

FARM TRANSPORT

FARM materials handling, or transportation, according to whether the new or old definition is preferred, has been anything but a spectacular success area. Until the fork lift reached the farm in recent years, we had really not advanced more than one step beyond the wagon and tumbril of the 1920s and 1930s. Inevitably, it was a case of removing the horse from the shafts and replacing it with the tractor.

True enough, the shafts gave way to the drawbar, but what could have developed into a revolutionary re-appraisal turned into a damp squib. A successful policy of shifting materials rests simply on moving a requisite to its final resting place in one movement. Failing that, then as few movements as possible should be made. And as it costs money to move things, unless strictly necessary they should not be moved at all.

BUILDINGS IN THE WRONG PLACE!

One of the great failings amongst farmers, yet one of the benefits to sellers of machines for produce handling, is the built-in habit that many farmers have of always siting a new building in the wrong place. Not so much wrong, perhaps, as inconvenient. The building can look right, blend with the environment and for most purposes it is right. Right until some person has to start unloading the bales, or sacks, or some other material. What is convenient to put produce in may not be so functional when the same produce has to be got out. The getting out is the part that is often forgotten when the building is planned. If planned is the right word!

It is amazing how far Dutch barns can be sited from the piggeries or fattening yards, for example. I have often thought that the answer to farm buildings planning is an old aircraft hangar.

The farmer should then be told that he can put things where he likes provided nothing spills out. Apart from the fire risk, I can think of no place better to plan one's farmstead than round a hangar.

It is unforgivable that one of the great inventions connected with farm transportation was not standardised. Even now, neither farmer nor machinery manufacturer has fully appreciated Harry Ferguson's invention of the pick-up hook. Nor have those concerned with farm safety learnt the wheelbarrow lesson to the full.

The wheelbarrow was the primitive farm tool used by Ferguson to demonstrate weight transfer to his confederates. A wheel at one end and handles at the other sufficed. Pick up the handles and you find the main part of the load weight is on your own essential moving part, the legs. Move the legs and the load moves. Bog down the wheel and the legs may still pull the load through. One appreciates that the pick-up hook could only come in the wake of hydraulics, but it was an integral part of farming's great invention. The hydraulics have to stand first, but farmers and engineers were so mesmerised by the sudden progress once the little grey Ferguson tractor went into mass production in late 1946 that an important aid was almost overlooked.

TRACTOR TRAILER COMBINATION

Farm transport is still mainly the rather untidy combination of tractor and trailer. None too manoeuvrable, not a very elegant combination, built-in danger factors galore, but functional and relatively sheap. So often, life and limb have to be weighed in fiscal terms against the alternatives. Certainly this happens in the sphere of transportation, whether by road or farm track.

Recently we have got around to worrying about the amount of downward thrust that is put onto the tractor drawbar, or link arms if the pick-up hook is used. As I indicated, my belief is that the drawbar part should be irrelevant, because we ought really to have been standardised on the hook long, long ago. With the wheelbarrow we had the safety factor of a man not picking up more than he was capable of lifting. Not so with the tractor. Hydraulics have become sophisticated and powerful.

Safety does not finish with picking up more than is good for the hook or drawbar in transit. Trailer makers are remiss in not making more use of reflectors at the rear, on the sides and even at the front end, too. Reflectors are the next best thing to full lighting.

FITTING FULL LIGHTING BOARD

Local councils and other authorities, besides some enlightened farmers, now tend towards the trailer lighting board as an accessory. Complete, ready-to-fit boards, with rear and stop lights, flashers, reflector triangles, number plate and anything else necessary are now on the market. For example, the Ward tractor lighting pack enables a farmer to fit his tractor with flashing indicators, a 7-pin socket and rotating flashing beacons. Wards, who introduced their trailer package in 1972, have had approval of their kits from the Design Centre and the boards with their two-point fixing can, of course, be switched from one trailer to another.

Many trailer making firms do offer the full lighting kit as an extra, but the elementary precaution of reflectors could so easily be an integral part of the trailer and thus go some way towards safety during the last journey home in late afternoon on winter evenings. As tailgates and even sides of a trailer are often removed, obviously reflectors would have to be mounted on the frame or floor section.

We hear a lot about trailer brakes but far less about illuminating them for road use. I cannot remember any trailer where the brakes worked for very long. Either they get used with the brake on, or they are full of mud, or maybe they are not at all that good in the first instance.

I am not so certain that we are not putting the cart before the horse with trailer brakes anyway. It could be just as important to see that tractor brakes are working efficiently. Since joining the EEC, it has become more necessary to look at the laws relating to the road use of agricultural vehicles on safety grounds, besides some re-appraisal of technology involved.

Jack-knifing with articulated vehicle combinations on our roads is as familiar as it is tragic. It can also happen on the farm and sometimes does, especially with tipping trailers, where bridging of the load can up-end the tractor. Some manufacturers have tried making trailers that are wider at the rear than at the front. This helps a lot when unloading materials prone to bridging, but there is a limit to how far one can go along those lines.

THE LAND-ROVER

The only real success story in farm transportation concerns carrying the farmer himself. The Land-Rover is used more for the specific purpose of transporting the farmer than for any movement of agricultural produce. The petrol engine can be four or six cylinders, developing 70·5 bhp, or 86 bhp, both DIN. The four-cylinder diesel engine is rated 62 bhp at 4000 rpm. My exprience is that the four-cylinder engine is adequate, although I have always found the petrol engine around 10 mph faster at top speed than the diesel.

I maintain that it should be possible to run any farm without the conventional farm trailer. Vehicles with four-wheel drive and of

81 The Land-Rover has four-wheel drive, rigid box-section construction chassis, non-corrodible light alloy body panels and galvanised external steel fittings. Available with 2286 cc 4-cylinder petrol or 2286 cc 4-cylinder diesel engines.

82 Steyr-Puch Halflinger with glass-fibre driving cab.

rugged construction, or as a proprietary type the forward control Land-Rover (particularly if it had a much larger pay-load) will probably negotiate tracks during the winter that the old tractor and four-wheel trailer would not.

Even the two-wheel, weight-transfer trailer of modern design, behind a two-wheel drive tractor, does not always get over the ground as one would wish. The arrival of Massey Ferguson's pressure control hitch did little to revive the flagging fortunes of the four-wheel trailer or the many that were made from old lorries in the early days after the waggon and tumbril went out.

OLD ROAD TRUCKS

In the main, old trucks are used by grain-growing farmers for carting corn in from the combine. As they often have tipping gears, they are ideal for this purpose. In fact, unorthodox or not, there must be a few cheaper ways; speed, that essential ingredient at harvest, is so much higher than with a tractor and trailer combination. Massey Ferguson, to their credit, did what they could to publicise the pressure control hitch which, by simple chain fixing to the trailer drawbar, pivots weight onto the tractor end. Farmers can use the hitch for other implements, too, if they wish.

For transporting sugarbeet and late-lifted potatoes, the old lorry is of little use, unless it is 4WD. Often, in bad seasons, one can see 4WD tractors struggling with two-wheel trailers. Something else that gets tried out from time to time is a power drive to the trailer axle.

WHOLE IDEA—TO REDUCE DRAUGHT

One idea, which apparently aroused more interest in Scotland than England, is to power the trailer axle from the tractor pto, put on traction types and disconnect the drive when it is not necessary. This is easy with the modern tractor with hydraulic engagement of the pto. The whole idea of farm transportation, it must be remembered, because of the mud, slush and slime encountered, is to reduce draught. Think back to the old tumbrils and waggons and it is easy to see that, with the limited power of the draught horse, rolling resistance was surmounted with large wheels.

The bigger the diameter, the less rolling resistance and the wider the surface, the less sinkage. But wheelwrights paid scant attention to the latter factor, if I remember aright! There are various difficulties with over-large wheels unless we go back to the waggon configuration

83 The Rekord all-purpose vehicle available with different bodies.

84 An example of a combination of different manufacturers' machines involving the forward control Land-Rover, a New Holland forager and a Farmhand body.

of wheels outside the platform. Similarly, it would have to be a delicate calculation and good gear selection if we went on to pto trailers. If the trailer wheels were being forced round at speeds much higher than the tractor wheels, obviously scrubbing and excessive tyre wear would take place. It is not so much drive as a little gentle assistance that is needed in muddy conditions, and the trailer wheels need not be going even as fast as the tractor wheels.

Twin wheels on trailers, conventional fashion, were quickly and firmly castigated with those early conversions from road vehicles. Rolling resistance, in a sugarbeet field in December, with a few square metres of slush rolling ahead of the twin wheels, proved the point admirably! As against that, the load on a pto-driven trailer could be more than doubled for the same power consumption.

Running true to form, we in Britain have largely taken the middle-course way out. We need more tyre area for flotation and weight carrying, so we overcome rolling resistance in the same way that the front wheel of a tractor helps the rear. We put double wheels in tandem.

CONCENTRATE ON TRACTOR TRACTION

By weight transfer, with the trailer wheels at the rear of the trailer, we have made it more logical that we should concentrate on tractor traction. In that we concern ourselves only with the rolling resistance at the trailer wheels and select a tyre section and a ply rating that, along with the axle strength, will sustain the load.

All the large-scale trailer manufacturers now go for tandem wheel layout on the bigger models, and this is serving the farming community reasonably well. Another facet of trailer manufacture that failed to stir the farmers was the three-way tipper. There are a few around and for specialised purposes they are useful.

Like the three-way tipper, the high-lift trailer is rather expensive. Yet I think that the latter is under-exploited. Using a scissor linkage connected to the chassis, the whole body is projected upwards about three metres, then tipped. For clamping potatoes or sugarbeet, the high-lift is superb. Unfortunately, the financial repercussion is a 100 per cent or more increase over the basic model.

TWO-WHEEL TIPPING TRAILER

By modern standards, I think it fair to say that the two-wheel weight-transfer, tipping trailer is a fairly simple piece of farm equipment. I know that rubber hydraulic rams and a mixture of metal

85 The single axle chassis of the Krone Agro-Trailer, shown here with the trailer body.

86 Carnfield 'La Baladeuse' high-lift trailer has a tipping height of 4 metres and a carrying capacity of 7/8 tonnes.

and wood is required and that all these things have had sharp price increases. This situation unfortunately has been an influencing factor in the move to secondhand trucks on the farm. Even for livestock feeding, and with plenty of mobile, self-discharging feed waggons on the market, it is noticeable that many farmers purchase old bulk-delivery lorries from the corn merchants.

Both the lorry with blower incorporated and the high-lift trailer are eminently suitable for bulk transport of seed and fertiliser direct to seeding or spreading machines.

One does not see them used for this purpose as much as one would have thought some years back. The advantages of packages for calibration purposes weights heavily against it. The same sort of remarks can apply to transporters for drills, crawler tractors and so on.

MARKET FOR SELF-PROPELLED TRANSPORTER

It has always been my view that there is a need in agriculture for the self-propelled transporter. Like the old lorries, the use of so many ex-industrial dump trucks rather illustrates the point. There is a market and an opening for this type of machinery at the right price.

For potatoes, sugarbeet and even grain in wet harvests, there is scope for the land-to-road transporter with its own propulsion unit. One assumes that such a machine would price itself out of the market before it got going. The 4WD vehicle, with a payload of anything between 10 and 20 tonnes, that could drive alongside the beet or potato harvester, then drive to store or factory (possibly with the chore of some cleaning of mud from the wheel) must therefore remain a dream for the present.

MINI CONTAINER SYSTEM

Another method of transportation that has been tried is the mini container system. Machines that pick up, transport and unload containers are mainly for the fruit fields. For general agricultural use, such machines as the Power X or Telecon, and the Krone from Germany, have dipped into the chilly waters of unorthodox farm transport. What would be orthodox at the docks or in the factory may not be acceptable on the farm.

With the Power X or Telecon type that picks up a body onto a common chassis, then puts it down where it is wanted and picks up something else, there is always the problem of ground conditions. Farmers cannot lay concrete everywhere and manoeuvring in mud, even to pick up a trailer with the pick-up hook, can be a hazardous job on the farm.

87 A trailer constructed for the bulk handling of seed and fertiliser from store to drill.

88 1-metre folding forks are standard on the McConnel Tail Slave TU20 fork lift.

89 Three big round bales can be carried at once with this conversion of the standard Sanderson Power Fork. The standard tines are replaced 1.5 metre tines.

90 The McConnel loader shown here can be driven from the tractor hydraulic system to load the Howard Rotaspreader, leaving the drawbar and pto free.

91 Built on the County forward control tractor, this self-loading lime spreader is fitted with a McConnel Power Arm 44 loader.

The manure spreader, a tool that transports and deals with its load at the end of the journey, is now of the most successful of transporters. Yet at the outset it relied on another great farming success, the front-end loader, for the commencement of its job.

Rear loaders seem to have come into the fray for no good reason, except, as always, for a specialised use. Fort-lift trucks, like the dump trucks, came in mainly secondhand because only the really big units can afford such aids. One of handiest of the rear-end type is the McConnel's Tail Slave, which has a rear fork lift and can have a built-in pick-up hook. The forks fold so as to avoid the trailer when it is picked up and moved.

SELF-EMPTYING TRAILERS

The self-emptying trailers are another good means of transport. It is quite remarkable how pick-up waggons are to be seen all over the Continent, yet how to slow UK farmers were to adopt them even in a limited way. Perhaps it is the odd experience of loading hay behind one of those machines with a rather over-anxious tractor driver that has put us off!

The slewing loader mounted on the tractor looked like being a success as it was able to perform standing still. However, use of concrete for sugarbeet loading pads helped to retain the popularity of the front-end loader. Yet the slew loader has had a certain amount of success in the field of self-propelled transportation. Using such aids as the ubiquitous McConnel Power Arm for fertiliser, lime or manure, it is not too difficult to assemble a self-propelled transporter that will load, transport and spread.

Using a tractor like the forward control County, such an outfit becomes self-propelled, or the versatility of a tractor-trailer combination can be retained with a Power Arm affixed to the tractor linkage and manure spreader behind.

CONTAINERISATION

Stillages, pallets and anything that panders towards the idea of containerisation have become very popular. With either a tractor-mounted fork lift, or the more dexterous self-contained type, a farmer can have his fertiliser delivered on pallets. Almost all the fertiliser firms now offer a pallet service.

While such firms as Fisons, ICI and Shellstar have adopted pallets, Pertwee try packages of several kilos, using a loader or hoist to sling them into position. By tipping over the spreader, of course, the load can be emptied straight into the hopper. Hoists, loaders and cranes have come into farming in various forms, self-contained or mounted on the farm tractor. Perhaps one of the most convenient methods is pneumatic conveying. Just create a high velocity air stream, and the more power that there is available from either tractor or stationary engine, the more material one can move.

The limitations are few, in that the materials do not have to travel in a straight line. Bends are no great problem except for wear on the ducting at corners with coarse materials (in case it is not generally known, malsters do not take kindly to malting barley that has been skinned by pneumatic conveying). The common grain auger is one of the most useful tools on the farm, in that it is both good and relatively cheap. However, since its inception, the problem of moving materials round corners has had to be looked into mainly for livestock farmers who use augers in feeding systems. Again, we do get some limitation, in that beans and peas will jam the flight in next to no time.

HYDRAULICALLY-OPERATED FRONT-END LOADER

Still no farm transporter usurps the primacy of the hydraulically-operated front-end loader. By the same token, nothing transcends the Land-Rover as the prima donna in transporting the boss around his acres.

Such things as automatic tailgates, spouts, grain sides, silage sides, high tippers and so on are usually available. As trailers have become larger, wood has very much given way to steel. For years trailer size lagged behind the power available, but now that we talk in the 10 to 15 tonnes bracket this is no longer so. Aluminium bodies to assist unloading, besides alleviating the risk of jack-knifing, are available on the Salop trailers.

Another attempt to tempt the farmer into an integrated transport system came with the Reform Multi Tractor. This little tractor, that looks more like a lorry, has available tipping body, a manure spreader, forage trailer, slurry tanker and sprayer. Not only has this tractor got four-wheel drive, but also a sleeve joint to allow axle oscillation.

THE WEIGHBRIDGE

Another connection with transportation is the weighbridge. We are now very much into the bulk age. The fork-lift truck comes into its own because of the need to calibrate with feed, fertiliser and seed, but a farm weighbridge suits loads of grain, either going in or out, much better.

The 'Weylode', a trailer attachment which weighs the load on tipping, is a very useful adjunct for the farmer wanting to know individual field yields; it is accurate to within 2 per cent, plus or minus. This accessory can also be used in conjunction with tractor loaders, transport boxes and, of course, tipper lorries. A portable pressure sensitive instrument with dial and quick release coupling is connected hydraulically. To weigh a load, the operator plugs the unit into the hydraulic coupling line, though a calibration chart is also needed after taking the dial reading. Calibration charts, as often used to get the reading on moisture meters, balers and seeders, always contain a nuisance value. It is so much easier to have an instrument that gives a direct reading. Unfortunately this is another case of price in relation to the market being a predominant factor.

SUMMARY

To summarise, the ultimate in farm transport excepting pneumatic conveying, pallets, stillages and containers has to be the land or road four-wheel drive lorry. Manoeuvrability and economy are important, so the vehicle has to be diesel, not petrol-engined—unless, that is, some newer power such as chargeable batteries (milk delivery vehicle style) can be found to be adequate. This is hardly a proposition except for very small transporters and fork lift trucks.

There is no shortage of trailer makers, and many of the large firms now have contract arrangements so that if one insists on a matching colour for the farm fleet of tractors, there is no problem be it the Massey Ferguson colour, Ransomes (Ford), International, Bamford and others.

I like the Salop trailers, then the Weeks or the Massey Ferguson which are identical. But if the farmer does not have to study the money involved—and there are few about—such Continental trailers as the French Brimont are so much more complete than the English versions.

Chapter 9

BUYING MACHINERY AND TRAINING THE OPERATOR

FARMERS are good, logical spenders. Some five-eighths of their net income is likely to be re-invested and around 20 per cent of each year's net income is going to be spent on machinery. It is hardly surprising that a good many salesmen are going to be knocking at doors for their own small part of some £1,600 million that is annually spent on farm requisites in the UK.

We incline towards the view that financially we are hard-pressed compared with our European counterparts, but the 1928 Agricultural Credits Act did enable the formation of the Agricultural Mortgage Corporation. Farming was thus able to get long-term credit up to two-thirds of the equity value of the holding. The relevancy of this to purchasing machinery is that farmers are able to mortgage two-thirds of the value of the farm if they want liquidity.

By the same token, AMC can be used for expansion of the farm. This is more likely to be the case, although there could be the problem of the inherited farm which had no mortgage, with cash badly needed to buy up-to-date machinery. A few years ago it would not have been sensible to talk of land values, mortagages and AMC in relation to buying equipment to run the farm. This is no longer the case. Machinery is now big money and many farmers now pay out a larger sum for a tractor or combine harvester than their holding cost them in the first instance.

LEASE-BACKS

If a farmer does not particularly want to own everything under his feet, owner-occupiers can turn themselves into tenants and thus obtain liquidity of money in the lease-back market. At the time of writing this book (May 1977) with interest rates (for borrowing) in the 12 to 15 per cent bracket, lease-backs could still be negotiated at around 5 to 6 per cent.

If one has not got the ready cash, the next best thing is a bank overdraft. In fact, overdraft money from one of the joint stock banks is probably always going to be the best way of buying anything, short of subsidised loan schemes, that is.

So we start off hoping that we can raise money out of trading profits to finance machinery purchase. After that, there is a good chance that the contractor is the best bet.

THE FARM CONTRACTOR

We could be on the verge of a greater acceptance of the agricultural contractor. Whether for whole-farm contracting or helping out to prevent over-capitalisation in machinery, the contractor can be the difference between being on top of the work on the farm and being in a gigantic muddle.

But it must be said that the old-fashioned conception of farm contracting is one of the worst features of mechanical farming. Its days are numbered. Planning for part-machine ownership and part-contract work, according to the economics of machine utilisation, has to take its place. For the first time since the advent of the mechanical age in farming, there are real prospects for the man who can bring farm contracting up to twentieth century standards.

Hire purchase, leasing, contact hire and any type of conditional sale have all to be carefully studied. Leasing stands apart from the others because the machines do not pass into the farmer's ownership, unless there is a clause allowing such purchase, possibly at the end of the stated leasing period. It will be recognised that where the ownership passes to the farmer he has the advantage of the taxation allowances.

There is now the added complication of Value Added Tax being added to rentals. This involves an interest-on-money factor between the times of paying out and reclaiming the money from Customs and Excise, though nobody would insinuate that this is other than a small net sum. It also depends on whether the farmer claims back by the one-month or the three-month period. Nobody would borrow long term for machinery purchase, but banks will negotiate a sort of medium-term credit that might be had at something a little above overdraft rates.

Farmers, as I have said, will not be on 'blue chip' rates, but will be paying around $2\frac{1}{2}$ to 3 per cent above base rate. Even assuming medium-term credit can be had at 3 to 5 per cent above base rate, this could still be a good way of buying machines that otherwise could not be obtained.

Apart from contract hire, there is often the opportunity to do short-term hire. Many agricultural dealers, finding themselves with combines on their hands at the outset of harvest or spare tractors at the commencement of spring seeding or autumn cultivations, will hire the machines out to farmers. Assuming the sale prospects for the immediate future are bleak and allowing for the machines coming back in 'off-farm' condition, it does reduce the stand-in book value by the amount gained from the hiring out. It suits both parties well, in fact. The farmer has had the short-term hire machine only for the weeks that he needed it and the dealer has recouped some of the 'dead' money.

Any farmer who is contemplating the pros and cons of machinery purchase really has no option open to him other than his accountant's advice. Normally, no accountant or bank manager would be brought into machinery purchasing preliminaries. The farmer needs to make those decisions within the farm gate. It is not a bad dictum to make a rule that the least people involved the better.

If re-investment in a good year avoids paying some percentage of the income tax bill and brings the farmer's stock-in-trade up to date, then every farmer must surely know that this is good business.

NFU-SPONSORED SYNDICATE SCHEMES

Now that they have the advantage of NFU-sponsored syndicate schemes, as well as other syndication arrangements through farmers trading organisations, there are many alternatives for machinery purchase. There are even syndication arrangements that can include the dealer. Most important is the fact that the farmer can probably get syndicate terms that are better than all the other arrangements. He may get down to a figure of 2 per cent over base rate which seems to be the low these days despite that 1 per cent over the old bank rate was a common denominator not so many years ago.

Where the dealer is involved in syndicating machinery, it is likely that a take-in is involved. Assuming a deposit of 20 per cent and the remaining 80 per cent paid off over the next five years or so, the farmer may not be parting with any hard cash initially because the deposit may well be covered by the trade-in. The dealer is only a sleeping partner, his main function being a consenting party for taking in secondhand trade-ins initially; there is also an obligation to buy back should the farmer default on his payments. As in all such syndication schemes, there has to be the inevitable charge to cover administration. This is usually 1 per cent of the cost of the new machine, spread over the repayment period.

SECONDHAND MACHINERY PURCHASE

Buying machinery from profits and keeping one's stock-in-trade up to date has to be priority number one, even if not always attainable. The farmer-dealer syndicate schemes, with their roots in the fenlands of Lincolnshire, can also be used for secondhand machinery purchase if necessary.

By and large, the axiom of keeping one man to his tractor and virtually one man to a machine serves the average farmer well. The answer at the tractor end of the argument is in favour of more standardisation on the individual farm.

If all the farm tractors are the same model, or from the same maker, it does obviate much of the inconvenience arising if the regular operator is ill or on holiday. In an emergency, a man can be switched to a tractor that would be identical to his own, or, at the least, very similar. Incidentally, I mention holidays rather with tongue in cheek because one assumes holidays are something to be planned for slack periods on the arable farm. Switching drivers from one machine to another because of holidays should not happen too often. Many farmworkers still prefer to take their holidays on odd, wet days. Normally, this is a mood that the arable farmer does not grumble about.

During the quarter century from 1952 to 1977 it is a salutary thought that a million people have gone from farming. Incredibly, where farmworkers were about 5 per cent of the working population 25 years ago, they are now down to half that proportion. When a farm had ten men and one fell sick, it was just 10 per cent of the farm's workers to be replaced, either with outside help or by a rearrangement of work allocation to the other nine. On a similar farm today, if a man is absent for any reason, he would more likely be 25 per cent of the workforce and possibly two-thirds. A 1000-acre, easy-working, light-land, arable farm can be managed by the boss and two tractor drivers, so that an absentee means a 50 per cent reduction in the hired labour.

LABOUR PROBLEMS

There can be no doubt about the problems that a low labour agriculture has brought to the individual farm. Such exigencies are a pronounced feature of farm life, as against the equivalent situation in an urban industry, for the simple reason of the farmer's requirements being a matter of timeliness. Such timeliness depends more or less entirely on the weather factor.

One hopes that there will always be a variation of farm size in the UK, brought about mainly by a retention of free enterprise. That way workers can move to different types and sizes of farms. A high-labour-content farm, such as one growing quick-freeze vegetables, with also livestock enterprises, can absorb a variety of types and ages.

Now that the service cottage has been virtually eliminated as a well-proven aid to labour mobility, most of the difficulties concerning farm machine operators will be augmented as time goes on. Thankfully, there is no need for the machine operator to live on the farm, so that the loss of the service cottage advantages to the farmer are less of a disaster than in the case of the livestock farmer. It is easy to dismiss the subject by referring to the fact that not many school leavers are interested in coming onto farms. Thus our labour force is getting older and older. This can hardly be a good thing.

However unkind, it has to be said that there is often no place for the boy on the all-mechanical farm. Yet farmers cannot expect to pluck staff from some hypothetical tree, fully trained, at the age of 25. The colleges and agricultural institutes of Britain are well known, but they are more for those in the management category rather than for machine operators. Ideally, farming should be prosperous enough for the farmer to be able to afford the odd learner, or apprenticeship boy, over and above the basic requirements in machine operators. That way the trainee, in late teens, is possibly good enough to take over from some older person who retires or moves on to less exacting, more sedate work.

TRAINING COURSES

Probably the courses now offered by farm institutes, mainly the day release courses, offer some solution to machinery operators' training. Unfortunately apprenticeship schemes have not been a great success. But it should not be forgotten that future machine operators will need specific training if the growth in machine sophistication is to keep pace with industry.

The trouble with day release courses in relation to the arable farm is that it may be the only fine day in the week! It is an excellent way of keeping a young man interested and his knowledge increasing all the time by means of four days on the farm and one day in the classroom. It is a good antidote to boredom. Yet many arable farmers will find the method of continuous block courses arranged during the off-peak period are as good. Regardless of how the knowledge is accumulated, whether day release course or tabloid course, machine operators do now have the advantage of some monetary reward which takes them nearer to the coveted high wages of their industrial counterparts.

There is now some sort of workers' ladder, with the categories designated Craftsmen, Appointment Grade Two, Appointment Grade One, Qualified Former Apprentices and Qualified Former Trainees. These are all steps on the ladder from the basic rate. So there is some incentive for training in agriculture, which, we must admit, was missing for a long time. Proficiency tests are drafted by panels made up of representatives from the industry. New tests get developed, validated and finally accepted (if satisfactory, that is) by the Agricultural Wages Board for inclusion in the Agricultural Wages Structure Scheme. As the examiner will have the power to select which tasks he wishes the candidate to perform within the defined category, the latter will need to be trained in all the various options. As all the main bodies concerned with practical farming get involved in proficiency testing, there is unlikely to be any great variation of principles or violent disagreement.

COURSES RUN BY MANUFACTURERS

Farmers should not overlook the efforts made by the manufacturers of farm machinery in this field. The Massey Ferguson training school quickly established a reputation for training men in the use of that firm's machinery, besides teaching them the rudiments of the engineering involved. Manns of Saxham, with their courses for Claas combine operators (especially the first-time owner), have sent both men and owner-drivers back to the farm elated by the interest shown in them.

If such courses awaken the operator to an interest in the machine under his command, that is one thing. If it reaches him to interest himself in the instruction book, that is another. Above all else, if operators are taught simply not to lose the instruction book, it is a worthwhile exercise! If farmers are taught not to lose them, it is equally good!

Complicated instruction books can be as confusing to the driver as they are useless to the farmer. Farm operators need the instructions to be given in print simply and concisely, as a sort of synopsis or in brief syllabus form. It is not common knowledge that at least one service does exist for careless machine operators who have mislaid the instruction book and are not afraid to tell their superior. David Brown is one company that attaches some amount of importance to the facilities offered by Reading University. As the facilities stretch back as far as the Ferguson-Brown models, this is no surprise. Using David Brown as the typical example of the machine maker whose after-sales service embraces the instruction book, they have deposited a master set of publications with the Museum of English Rural Life at Reading University.

The depositing of such archival material by manufacturers means that they can direct enquiries to Reading, who provide a good photocopying service. So much importance has been attached to instruction books in the area where so much field machinery is used that in 1972 the East Anglian branch of the Institution of Agricultural Engineers arranged a study group and subsequently produced a Writers Guide for the instruction of anyone engaged in compiling machinery instruction books.

INFORMATION ON MICROFICHE

The Agricultural Engineers Association and the British Agricultural and Garden Machinery Association have agreed a common specification for the presentation of technical information on microfiche. David Brown Tractors researched the project and brought the myriad technical details together into a concise but comprehensive specification. The microfiche system is the reproduction on film of multi-page publications, such as service manuals and parts catalogues, and their projection onto a screen. Over 400 pages can be accommodated on a single fiche. The advantages are that there is no need for bulky publications to be distributed to the dealers; fiche can be produced more quickly, and modifications and updating able to be done faster. More important, it is cheaper.

The adoption of a standard for the UK should ensure that all fiche produced by most of the manufacturers can be used in the same type of viewer. This is important. Farmers have suffered enough from a lack of standards in the machines themselves, without machinery agents bringing any more complication into an already complicated existence. All this may involve the machine operator far more in the years to come than has been the case hitherto.

VIDEO DISPLAY UNIT

Before concerning ourselves with the farm operators' niche in all this, it is as well to refer to the effort made by BAGMA to illustrate the latest in technological innovation in stock control and parts identification. A video display unit has been exhibited in operation giving dealers a complete breakdown of stock availability, orders outstanding, sales history and pricing details, displayed on a screen within seconds of keying in the part number. Having now reached agreement regarding the microfiche reader mentioned earlier, BAGMA and AEA lost no time in recommending the Caps Alpha Reader. Caps is a UK company.

Except for farm managers, foremen and farm mechanics, the spares (or lack of them) has been the farmers' prerogative. For the most part, it is going to remain that way, but not exclusively so. To get value for money out of the large, sophisticated machines that are now entering agriculture, it may well be necessary to short circuit the collection of urgent spares. The hours lost when an operator has to go looking for the boss to report a breakdown will have to be looked at as the machines get bigger and more expensive.

TRACTOR-CAB COMFORT

Many manufacturers busy themselves with supreme operator comfort in the form of radio sets in the cab, but few pay much attention with private mobile radio (PMR). This could provide another dimension, indeed it does already, but usually only on very large farms. We have only touched the fringe of telecommunications so far. Once we have fully-trained operators using giant machines on the majority of farms, and especially in the sphere of contracting, the possibilities of radio links to machinery depots are tremendous. Already the link exists between farm office and worker, but this can fail hopelessly if the office is not always manned. Certainly it is unlikely that the office is going to be manned for all the hours that are worked outside during peak periods. So vital is the need for big machines to be kept going during the rush periods that the radio link between the tractor or combine driver and the depot that supplied the machine is a distinct possibility. Already dealers arrange shifts that take in Saturdays and Sundays, and, in some cases, 24-hour working during peaks such as harvest.

For the intensive farm, where, for example, co-operative gangs of machine operators are using the retinue that accompanies the harvesting of quick-free crops around a given area, a link between one man in a Land-Rover and an office would suffice. On the extensive grain-growing farm where a man will have a lonely existence on a giant machine in very large fields, there may be difficulties of communication. Pye Telecommunicatioas are well known for their transmitter-receiver handsets, besides their mobile radios for mounting in the tractor cab. If there is not room in the cab, a remote control model allows the control panel to go inside and the actual unit outside.

COMMUNICATION DIFFICULTIES

There is no doubting the fact of communication difficulties in prairie-type farming. One of the worst things is the simple puncture because the driver cannot propel himself back to the farmstead. With many breakages it is still possible to drive back to base and report.

Whether with a radio link to farm office, or to local depot, or with traditional means of communication, the machine operator in the coming era of giants will need to be given the necessary authority and means of transport to keep his machine in action. Provided operator are reliable, responsible and trustworthy, there is no harm in authorising them to use a farm vehicle (or their own with suitable reimbursement) to collect a spare part when the boss, manager, or overseer is not available.

CONTINENTAL MACHINERY CENTRES

We are not blessed with the sort of training schools for farmers and farmworkers that are to be found in some Continental countries. I have visited one of the superb machine centres in Germany, where it is impossible to simulate combine harvesting, but with a dirt floor; in these gigantic buildings winter courses can be organised and all tillage and seeding work simulated.

Tabloid courses for machines other than the prime mover are usually of a duration equal to the course on the tractor. This illustrates fairly the system we are committed to, namely, a prime mover with ancillary implements, excepting the few instances of self-propulsion. A pleasing aspect of such education is the willingness of machinery firms to supply machines for tuition and without too much concern about the cost. History of the German schools goes back to 1914, and when I was there as a guest of Krone (UK) Ltd in early 1972 there were fourteen such establishments.

TEACHING SAFETY ON THE FARM

With around 100 deaths a year in agriculture, the teaching of safety is now a worthwhile exercise. Much of our meagre workforce is injury-prone. If there is a safe and an unsafe way of doing a job—and there usually is—the latter will far too often be chosen. Getting interest in first-aid courses at the training establishments is never easy and there is no reason to suppose that pure and simple safety courses would be any more popular.

The only acceptable method and the one that is often being adopted, is to incorporate the safety element into all machinery courses. Sometimes the difference between safe and unsafe machinery usage is no more than the difference between the right way and the wrong way. It took a lot of heavy print in the early instruction books to warn operators against pulling from the top hitch point of a tractor. Now that safety cabs are law of the land, the chances of death from such elementary misdemeanours are less, but the ancillary losses from injured persons and damaged machines are as great.

Now that most arable farms are equipped with a workshop of some kind, it is as well to state that not all accidents concerning machinery happen in the field. The farm workshop also takes in the risk inherent in the untidy. Falls, blows and wounds account for the most number of accidents.

From the simple incident of the fork laid down instead of being stood up, to the electric current left switched on, absent-minded individuals will continue to be responsible for accidents to themselves and others. It is the job of Farm Safety Committees, besides all those engaged in educational courses, to get the message across in an effective manner.

Thus we conclude that instilling the art of self-preservation into machine operators is something that will be taught from habit, whenever and wherever machinery topics are discussed.

It should be no less evident in the courses organised by the machinery manufacturers and retailers than in those run by recognised training establishments. It is an equally good idea if the farmer sets his employees an example. Having financed the machine purchase and trained the operator, it is planned economy to keep them both fit and well.

Appendix

TABLE FOR CONVERTING BRITISH AND METRIC MEASUREMENTS

LENGTH

BRITISH TO METRIC		METRIC TO BRITISH	
1 inch (in)	= 2·54 cm or 25·4 mm	1 millimetre (mm)	= 0·0394 in
1 foot (ft)	= 0·30 m	1 centimetre (cm)	= 0·394 in
1 yard (yd)	= 0·91 m	1 metre (m)	= 1·09 yd
1 mile	= 1·61 km	1 kilometre (km)	= 0·621 miles

Conversion Factors

inches to cm	× 2·54	centimetres to in	× 0·394
or mm	× 25·4	millimetres to in	× 0·0394
feet to m	× 0·305	metres to ft	× 3·29
yards to m	× 0·914	metres to yd	× 1·09
miles to km	× 1·61	kilometres to miles	× 0·621

AREA

1 sq. inch (in^2)	= 6·45 cm^2	1 sq. centimetre (cm^2)	= 0·16 in^2
1 sq. foot (ft^2)	= 0·093 m^2	1 sq. metre (m^2)	= 1·20 yd^2
1 sq. yard (yd^2)	= 0·836 m^2	1 sq. metre (m^2)	= 10·8 ft^2
1 acre (ac)	= 4047 m^2 or 0·405 ha	1 hectare (ha)	= 2·47 ac

Conversion Factors

sq. feet to m^2	× 0·093	sq. metres to ft^2	× 10·8
sq. yards to m^2	× 0·836	sq. metres to yd^2	× 1·20
acres to ha	× 0·405	hectares to ac	× 2·47

VOLUME (LIQUID)

1 fluid ounce (1 fl oz) (0·05 pint)	= 28·4 ml	100 millilitres (ml or cc)	= 0·176 pints
1 pint	= 0·568 litres	1 litre	= 1·76 pints
1 gallon (gal)	= 4·55 litres	1 kilolitre (1000 litres)	= 220 gal

Conversion Factors

pints to litres	× 0·568	litre to pints	× 1·76
gallons to litres	× 4·55	litres to gallons	× 0·220

Table for Converting British and Metric Measurements *(Continued)*

WEIGHT

1 ounce (oz)	= 28·3 g	1 gram (g)	= 0·053 oz
1 pound (lb)	= 454 g or 0·454 kg	100 grams	= 3·53 oz
1 hundredweight (cwt)	= 50·8 kg	1 kilogram (kg)	= 2·20 lb
1 ton	= 1016 kg or 1·016 t	1 tonne (t)	= 2204 lb or 0·984 ton

Conversion Factors

ounces to g	× 28·3	grams to oz	× 0·0353
pounds to g	× 454	grams to lb	× 0·00220
pounds to kg	× 0·454	kilograms to lb	× 2·20
hundredweights to kg	× 50·8	kilograms to cwt	× 0·020
hundredweights to t	× 0·0508	tonnes to tons	× 0·984
tons to kg	× 1016·0		
tons to t	× 1·016		

INDEX

After-sales, tractors, 26
Agents, reliability, 34
Agricultural Mortgage Corporation, 200
Agrostroj, sugarbeet pick-up waggon, 163, 172
Air-cooled tractors, 32
Aircraft spraying, 106
Allaeys, self-propelled sprayers, 105,106
Allis-Chalmer's combine harvesters, 116, 119
—, —, gleaner, 127
—, —, —, Hillside, combine harvester, 127
—, —, 60 model, 119
—, —, 5000 model, 127
—, electrically-powered tractor, 39
—, Roto-Baler, 134–135, 136
Allman's chemical self-filler attachment, 103, 106
Allman sprayers, 95, 106
—, —, 'Intaro', 103, 106
Alpha-Accord drill, 69
—, spinning disc spreaders, 84
—, Tornado, 60
Alvan Blanch broadcaster box, 81
Amac root planter and harvester, 90, 172, 184
Amazone one-pass drill-cultivator-roller, 80, 91
Anti-freeze, tractors, 31–2
Aquatic Weed Control Boat, 102
Armer-Salmon Universal cleaner-loader, 166
Atkinson's Hydramatic, 84, 91
Automatic gear-boxes, 35–7
Autumns, good and bad, planting methods, 45–46
Avadex wild oat control, 98

Bales and baling, badly lodged crop, 140–141
—, effects of climate, 155–156
—, endless chain pick-up conveyors, 149–151
—, random loading, 148–149
—, round, 134–135, 141–142
—, sizes and shapes and uses, 142–146
—, sledge, 146–148
—, trailers, 151–153
—, transport, 148–153
—, unrolling, 146
Baler(s), future, 155
—, low-density, 135–136
—, ram, 139
Bamford big baler, 157
Bamford-Kvernelands disc harrows, 56
Beans, harvesting, 115
Benedict Amac planter and harvester, 90, 172, 184
—, power harrow, 59
—, soilmaster, 59
Bèrthoud trailer sprayers, 103
Bettinson double-disc coulter, 69, 72
Big A self-propelled spreader, 91
Blowing better than rolling, 112–113
Boggy and fenland areas, potato harvesting, 183
Bogie tractors, 38–9
Bomford and Evershed range, 65
—, —, tillage equipment, 55
Boom height, 99–101
Boots sprayers, 95
Boughton tractor, 183
Brimont trailers, 199
British Sugar Corporation, 159
—, demonstrations, 160, 161, 163
Broadcaster boxes, 81
Broadwater band sprayers, 103
Brown, David, *see* David Brown
BTS-Pottinger, sugarbeet pick-up waggons, 163

Cantone; drill, till and fertiliser machine, 77

Carier drills, 69, 80, 81
Caruelle trailer sprayers, 103
Catchpole Cadets, 161, 162, 163
Caterpillars, 42, 56
Century Fabrications tractor for boggy areas, 183
Cereal, drilling, 72–77
—, *see also* Grains; Combine harvester
Chafer's liquid fertiliser conversion, 89
—, saddle tanks, 81, 103, 106
—, sprayers, 94, 95, 106
Chance's chisel plough, 60
Chemicals, modern uses, 45
Chisel ploughing, 45–46
—, ploughs, 60, 65
Claas combine harvesters, 119, 123, 124, 133
—, Retromat, 123, 129
—, Rollant baler, 142, 156
—, straw delay service, 131
Claey combine harvesters, 119
Clayson 8080 combine, 129, 131, 133
——, re-threshers, 127
Clevis hitch, 59
Combine harvester, 107–133
—, automatic height control, 124
—, automatic speed control, 128
—, background history, 107–111
—, blowing, 112–113
—, hilly areas, 127–128
—, justifying costs, 118
—, money difficulties, 110–111
—, round-and-round working, 132–133
—, self propelled, 118–124
—, serrated knives, 113–114
—, size and shape of cabs, 129–131
—, swathing, 112–113
—, tanker combine, 109–110
—, wrap-round or trailer types, 119
Communication difficulties, 207–208
Container system, mini, 193
Containerisation, 197
Contractors, 201–202
—, working hours and tractor use, 34
Cooper Octogang disc harrow, 59, 65
—, Tareaway cleaner-loader, 167
Coulters, 69
—, double-disc, 69
—, Suffolk, 69
County tractor, 34
Courses run by manufacturers, 205–206
Cousins levelling harrow, 60
Crawler tractors, 41–42
CTM forager-type top-saver, 163
—, sugarbeet cleaner-loaders, 166
Cultivator, rotary, 49–50, 80
—, —, adaptations, 50
—, —, Underbuster 50
—, spring-tine, 48
—, tine types, 56 . .
Cutter bars, 114, 115

David Brown agencies, 29
—, automatic gearbox, 35
—, hydra shift, 35–36
—, 2D tool carrier, 32
—, safety cabs, 36
—, tractors, popularity, 35
—, —, 990 model, 33
—, transmission systems, 36
Deere, John, *see* John Deere
Delvano, self-propelled sprayers, 105
Demonstrations, potatoes and sugarbeet, 160, 161, 163, 180–181
Desiccants, 117, 118
Deutz four-wheel drive tractor, 32
—, Intrac system, 32
Diesel compression engine introduction, 16
—, versus oil, 39
Diggers for potato harvest, 178, 179
Direct drilling, 45, 72
Disc harrows, 56
—, folding, 59, 65
—, John Deere, 18
Doggett (Engineering) Geith, stone picker, 90
Dorman band sprayers, 103, 106
—, saddle tanks, 103, 106
Dowdeswell tractor, 65
Drainage 63, 64
Drill-cultivator-roller, 80
Drill-and-till implement, 77, 80
Drilling, combine, 67–8
—, direct, 45, 72
—, optimum spacing, 72–73
—, precision, 73, 75
Drills, force-feed, 68–79
—, preferences, 91
—, seed and fertiliser, 66
—, tine, popularity, 69
—, types, 80

Evers & Wall 'Tween-Row', 103, 106

Farm contractor, 201–202
—, machinery, *see* Machinery
—, mechanisation, inefficiency, 29
Farmhand-Vermeer, round baler, 153, 156
Farming, starting, 24–6
Farmworkers, grades, 205
—, manpower reduction with tractors, 16–19
—, working hours and tractor use, 34
Fendt air-cooled diesel tool carrier, 32
Fenland and boggy areas, potato harvesting, 183–184
Ferguson, Harry, farm machinery pioneer, 16
—, pick-up hook, 186
—, TE20 tractor, 33
—, —, with combine harvester, 118, 119
Fertilisers, distribution, 81–89
—, drilling, 66, 67, 77, 81–83
—, —, injectors, 84–89
—, —, liquid, advantages and disadvantages, 81
Fertispread, 84, 91
Fiat crawler tractor, 42
—, sales points, 29
Fields, round working against straight lines, 44
Fisons, fertiliser spreading, 83–4
—, marker systems for spraying, 96
—, saddle tanks, 103
Force-feed drills, 68–69
Ford agents, 34
—, dealers and agents, 26
—, Henry, farm machinery pioneer, 16
Ford tractors, early, 19, 33
—, 5000 model, 33
—, —, 7000 model, 33
Fork-lift trucks, 196
Four-wheel drive, and potato harvesting, 183–184
—, tendency, 32
Fowler's Gyrotiller, 20, 24
Franquet pick-up loader, sugarbeet, 172
Freeman automatic bale waggon, 154
Front-end loader, hydraulically operated, 197–198
Furrow shapes, 53

Gear-boxes, tractor, 35–7
Geith, stone picker, 90
Grain, aeration, 116–117
—, moisture content, 116
Grain monitors, 124–127, 128
—, sealed silos, 115–116
—, storage after combine introduction, 111–12
—, from straw separation, 131
Grimme self-propelled potato harvesters, 182, 184
Gruse planter, 89
Gyrotiller, 20, 24

Hardi Pick-up Unit Sprayer, 102
Harrows, chain, 60
—, disc, 18, 56, 59, 65
—, folding, hydraulic, 59
—, power, 59–60, 65
—, rolling, 60
Harvesters, potato, 179, 180–184
—, sugarbeet, 38, 161, 162, 163, 172, 184
—, *see also* Combine harvesters
Harvesting roots, 158–184
Hassia automatic planter, 90
Haulm extractors, 182
Herriau sugar beet equipment, 172
Hesston, bale pick-up and compressor, 151, 157
Hilly areas, combining, 127–128
Hitches, 32
—, clevis, 59
Hook, pick-up, 186
Hoppers, individual, disadvantages, 76–77
Horse as draught animal, 15
Horsepower, farm needs in tractor size, 37–8
—, to manpower, changes, 16–18
Horstine, fertiliser placement and granule applicators, 81
Hovercraft, 39
Howard big baler, 141, 142
—, rectangular baler, 157
—, Rotadigger, 61, 65
—, Rotaplanter, 89
Huard ploughs, 55
Hunt broadcaster box, 81
—, rolls, 56
Hydramatic, 84, 91
Hydrostatic transmission, 37–7
—, and potato harvesting, 183–184

ICI Aquatic Weed Control Boat, 102
—, 519 minimum tillage drill, 80
Information on microfiche, 206
Instruction books, 205, 206
International Harvester, 35–6
—, balers, 135, 140, 157
—, combine harvesters, 119, 131, 133
—, —, trailed, 119
—, distribution, 29
—, drills, 66, 91
—, hydrostatic transmission, 36
—, 275 loader tractor, 33, 140
—, safety cabs, 35, 36
—, seed and fertiliser, drills, 66
—, 374 tractor, 35
—, transmission systems, 36

Jack-kniving of farm vehicles, 187
Jethro Tull stategy, 67
JF straw processing with caustic soda, 153
John Deere balers, 157
—, combine harvesters, 133
—, —, Sidehill 965H model, 127
—, Cross-Shaker, 131
—, sales bases, 29
Jones Invicta self-propelled baler, 135

Kidd rolls, 56
Knives, beet harvester, 173
—, serrated, 113–114
—, sharpeners, 139
—, twine, shear, mower, combine, 139
Kongskilde Triple K spring-tine cultivators, 46, 65
Krone mini containers, 193
—, ploughs, 55
—, sugar-beet pick-up waggons, 163
Kronedrill, 81
Kronevator, 65
Kuiken sugarbeet harvester, 172

Labour problems, 203–4
Land-Rover, 187–189, 198
Lawrence Edwards 'Quickie' sugarbeet loader, 167
—, roller vane pump, 102
Leaper Ramsay fertiliser injector, 84
Leasebacks, 200–201
Lemken equipment, 65
Lely's Roterra seed-bed maker, 20, 60–61, 65, 80, 91
—, Victory combine harvester, 133
Leyland's tractors, 36–37
Lilliston rolling cultivator, 59
Lighting board, 186
Lister Big Bale Spear, 153
—, chain harrow, 60
—, 'Take-Put', 149, 153
Lockwood Graders, self-propelled potato harvesters, 182
Loctronic precleaner, potato, 182
Lorant baler, 135, 136

Machinery, buying, 200
—, —, secondhand, 203
—, centres, 208
—, farm, pioneers, 16–18
—, instruction books, 205, 206
—, keeping to one or two manufacturers, 29–30
—, standardised, advantages, 30
Mann-Claas sugar-beet pick-up waggons, 163
Mann-RDA grain monitor, 126
Manpower from horsepower, changes, 16–18
Manufacturers, different, farming needs, 29–30
Manure spreader, 196
Marker system for spraying, 95–96, 103–6
Marshall's crawler tractors, 41
Massey-Harris 21 combine harvester, 119, 131
—, seed and fertiliser drills, 66
—, 1930 tractor, 20
Massey Ferguson 128 baler, 140, 157
Massey Ferguson combine harvesters, 133
—, —, 726 model, 124, 133
—, —, 760 model, 123–124
—, —, 780 model, 124
—, diamond plough, 53, 55
—, distributors and dealers, 26, 29, 30
—, '30' drill, 80, 91
—, '85' plough, 53
—, 711 potato harvester, 181
—, re-threshers, 127
—, 'Suretie', 142
—, 135 tractor, 33
—, —, 1200 model, 37, 38
—, —, 1505 model, 38

—, —, 135 and 500 models, 35
—, trailers, 199
—, transmission systems, 36
—, Waferer, 141
Matrot 6-row sugarbeet harvester, 172
McConnel Bale-Packer, 149
—, Power Arm, 197
—, —, with beet grab, 167
—, tail slave, 196
McCormick Harvester 35
Mechanaid, Scimitar, 65
Melotte sugarbeet pick-up waggons, 163
Mercedes Benz four-wheel drive tractor, 32
Mini container system, 193
MM G8 combine harvester, 199
Moore Uni-Drill, 72
Moreau 5-row complete beet harvester, 167, 184
—, —, with defoliator, 167, 172
Mortgages, 200
Muir-Hill 4WD tractor, 34, 41, 193
Multibeet, 162
Museum of English Rural Life, Reading, 206

Nelson Farm Equipment Company bale trailers, 151
New Idea, 39, 42
—, Uni system, 154
Nitrogen drilling, 73
Nodet drill, 80
Nordsten drills, 91
—, —, grass seed drill, 80
Nozzles, selecting, 99–101

Oliver combine harvester, 119
Otto engine, 16

Parminter power harrow, 59
Pearson Plantmasta, 89
—, stone picking machinery, 90–91
Peas, harvesting, 115
—, planters, 89–90
—, moment to cut, 117
Peasenhall drill, 68
Pettit disc plough and cultivator, 56, 65
Phosphates, 67
Pick-up hook, 186
—, reel, 114–115
—, Unit Sprayer, 102
Picking up bales behind combine, 140
Pitchpole harrow, 59
Plant Protection 'Vibrajet' nozzle system, 103
Planting depth, 73–74
Plantmasta, 89
Plough, disc, no support, 47
—, farming without, 64–5
—, one-way, 53–54
—, —, automation, 53–55
—, semi-mounted, 55–56
—, unbeatable, 50
Ploughing, bad, versus no-plough, 50
—, chisel, 45–6
—, deeper, for beet and potatoes, 46
—, drill-and-till era, 46
—, furrow shapes, 53
—, speed, effect, 53
—, technique, importance, 50–52
Pneumatic seeding, 85–89
Potatoes, farm-to-consumer selling, 179
—, harvesters, self-propulsion, 182
—, —, testing prototypes, 181–181
—, harvesting, 175–184
—, —, fenland and boggy areas, 183
—, —, ground pressure, 183
—, —, unwanted stones, 180, 181–182
—, planters, 89–90
—, ploughing for, 46
Power units, 39
Power X, 193
Pumps, spraying, 98–99

Radiators, anti-freeze, 31–2
Radio links between farm and tractor, 207–208
Ram baler, 139
Ransomes double drum thresher, 131
—, Faun, 90, 180, 184
—, Hunter, self-propelled harvester, 172, 184
—, ploughs, 33, 65
—, sprayers, 106
Ransomes-Nordsten Universal sowing unit, 80
Raussendorf press, 139
Rear loaders, 196
Reel, pickup, 115
—, setting, 115
Re-threshers, 127
Richard Pearson, *see* Pearson

Ritchie 'Transfer Weight' transporter, 151
Roadless tractor, 34
—, —, crawler tractor, 42
—, vane pump, 98–99, 102
Rolls, 56–8
—, coupling, 56, 58
—, flat, 56–8
—, ribbed-type, 56–8
La Ronde, 60
Root crops, harvesting, 158–184
Root Harvesters, Gruse, 89
—, Todd cleaner-loader, 166
—, X-ray and electronic potato-sorting system, 181
Rotadigger, 61
—, attachment, 50
Rotaplanter, 89
Rotary cultivator, 49–50
—, tiller, 60
Rotaspa spading machine, 60
Rotaspike attachment, 50
Roterra seed-bed maker, Lely's, 20, 24, 60–61, 65, 80, 91
Roto-threshing, 128–29
Round bales, 134–135, 141–142
Round-the-field working, 44
Row widths, and suitable machinery, 87–89
—, for sugarbeet, 87–89

Saddle tanks, 81, 103
Safety on the farm, teaching, 208–209
Salmon 21 self-propelled sugarbeet harvester, 172
Salop trailers, 199
Sampo rotary harrow, 59
Sands self-propelled sprayer, 105
Schilter four-wheel drive tractor, 32
Scimitar, 65
Seed covering, 76
—, quality, 74
—, uniformity, 74–75
Seeding, pneumatic, 85–89
Silos, sealed, 115–116
Sisis Equipment rotary cultivator, 80
Smallford planter, 90
Soil losses in sugarbeet harvest, 165, 166, 175, 178
Solobeet, 38, 172
Spare wheel as necesssity, 30
Sperry New Holland bale-handling systems, 149–151, 157
—, bale waggon, 151
—, double drum thresher, 131
—, TR 70 roto-thresher, 128
Spinners, 81–84
—, for potato harvest, 178
Spinning disc fertiliser spreaders, 81–84
—, care in use, 81–3
—, tray system, 83
Sprayers, 92–106
—, cost-savings, 93–94
—, mounted, 94–95
—, sizes, 102–3
—, trailer models, 103
Spraying by aircraft, 106
—, late, straw inhibitors, 117–118
—, marker system, 95–96
—, underlapping, 98
Spring-tine cultivator, 48
Standen Multibeet, 162, 163
—, Rapide, potato harvester, 184
—, Solobeet, 38, 172
—, —, Cyclone, 172
—, —, sugar-beet harvester, 38
—Startrite, 184
—, sugarbeet equipment, 38, 161, 163
—, sugarbeet topper unit, 173
—, tandem drill, 81
Stanhay belt feed drill, 85, 91
—, drill, 72
Steam power on farms, 16
Steiger Couger 250 hp tractor, 37
Stone pickers, 90–91
Stones, unwanted, in root harvests, 163, 180, 181–182
Strautmanns beet-top waggons, 163
Straw, balers, low-density, 135–136
—, burning, 134, 137, 138, 155
—, converting to animal food, 137–8
—, effect of climate, 155–156
—, harvest, 135–157
—, inhibitors, 117–118
—, permanent cover, 136–147
—, processing with caustic soda, 153
—, processing machines, 154–155
—, round bales, 134–135, 141–142
—treatment, 113–114
Sub-soilers, 60, 61, 64
—, trailed and mounted types, 64
Sugarbeet as break crop, 159–160
—, cleaners, 164

—, cleaner-loaders, 166
—, drilling, 72–77
—, early machinery developments, 161–162
—, forage toppers, 165–166, 173
—, future in Britain, 158–159
—, harvesters, 38, 161, 162, 172
—, —, continental, 162
—, on heavy land, 160
—, machinery in use, 159
—, pick-up waggons, 163
—, ploughing for, 46
—, revolving cage front loader, 167, 172
—, row widths, 87–89
—, seeding, 85
—, soil losses, 165, 166, 175, 178
—, sowing, 87
—, tankers, 162
—, topper unit, 173
—, tops as food, 162, 163
—, yields, 159
Swather, 112
Swathing, 112–113, 116–117
Syndicate schemes, NFU-sponsored, 202

Taarup, straw processing with caustic soda, 153
Tasker's Amazone, 80, 91
Tasker Paterson Fertispread, 84, 91
Telecommunications, 207–208
Telecon, 193
Tenderometer, 117
Tillage, no-plough techniques, 43–4
—, tools, power-driven, 48–49
Tine drills, popularity, 69
—, types, 56
Todd cleaner-loader, 166
Tool carrier, 32
Topper, flail, 174–175
—, units, 173
Topping knives, 174
—, overtopping or undertopping. 175
Tornado, 60
Track Marshalls, 41
Tractors, 16–42
—, agencies, 26, 29–30
—, air-cooled, 32
—, automatic, 33, 35–6
—, as basic equipment, 25, 26
—, best buys, 33–4
—, bogie, 38–9
—, buying, 26
—, —, weight considerations, 29
—, cab comfort, 207
—, carriers, 39
—, choosing, 34–5
—, cost, 19–20
—, —, compared with workers' wages, 19–20
—, crawler, 41–42
—, dealers, 26, 29–30
—, early types, 20
—, electrically-powered, 39
—, Ford 5000, cost, 19
—, Fordson, cost, 19
—, hitches, 37, 40
—, horsepower needs, 37–8
—, MF 1200 model, 18
—, MF 185 (Fergie)), 17
—, numbers used, 16–18
—, owner-operators, 18–19
—, preferences, 42
—, reliability, 26
—, trailer combination, 186
Trailer brakes, 187
—, lighting board, 186
—, self-emptying, 197
—, traction, 191
—, two-wheel tipping, 191–193
Training course, 204–205
Tramline technique, 85–87
Transporter, self-propelled, 193
Transportation, farm, 185–199
Triple K spring-tine cultivator, 65
Trucks, 189
Turbair Sprite, 102
Turbo-charging, 40–41
Turbotiller, 59
'Tween-Row', 103
Twine, 140
Tyres, pneumatic, benefit, 31
—, radial, performance 31
—, size, 31

Ugerlose front hoe, 103
Underbuster attachment, 50
Uni-system, 154–155

Valmet 1502 bogie tractor, 38
Vibrajet nozzle system, 103
Vicon/Alpha-Accord spinner, 69
Vicon finger wheel rake, 140
—, Gehlbale, 142, 153, 157

—, power harrow, 59, 65
—, Rotaspa spading machine, 60
—, spinning disc spreaders, pendulum, 84, 91
Video display unit, 206–207
Volvo beet loader, 167

Wafering, 141
Wages, farmworkers', and tractor costs, 19–20
Ward tractor lighting pack, 187
Warnock trailer sprayer, 103
Water ballasting of tyres, 31
—, tanks, mounted, 101–2
Webb floating ridge drill, 81, 85
Weed control, 92–106
Weeks trailers, 199
Weighbridge, 198
Welgar low-density baler, 135, 157
Weylode, 198
Wheels, cage, 31
—, on farm vehicles, draught, 190–191
—, spare, as necessity, 30
—, twin, 31
White tractors, 37
Whole-crop harvesting, 138–139
Wild oats, 94, 98
Wilders Cultharrow, 59

Zetor, sales points, 29